MODEL LOC

their desig

Members of the Chesterfield Society examine the 5 in. gauge "Firefly" built jointly by the author and Norman Spink. This was the first 5 in. gauge locomotive to be fitted with a thermic syphon

An example of a locomotive fitted with a short barrel boiler with wide firebox; built by Mrs. Y. Etter of South Africa

[*Frontispiece*

Model Locomotive Boilers

their design and construction

MARTIN EVANS

Editor of

MODEL ENGINEER

FOREWORD

by

Professor D. H. Chaddock, C.B.E.

MODEL & ALLIED PUBLICATIONS

Argus Books Ltd., Watford,

Hertfordshire, England

Model and Allied Publications
Argus Books Ltd.,
St. James Road, Watford, Herts, England

First Published 1969
Second Impression 1973
Revised Edition 1976

ISBN 0 05242 483 3

PRINTED IN GREAT BRITAIN BY HEADLEY BROTHERS LTD
109 KINGSWAY LONDON WC2B 6PX AND ASHFORD KENT

Contents

List of Plates

Foreword

THE reputation which steam locomotives, large and small, rightly enjoy for power and flexibility derives not so much from their engines as from their boilers. It is here, by the combustion of fuel, that chemical energy is transformed into thermal energy and again, by the evaporation of water into steam, that thermal energy is transformed to pressure energy which the engine can use. Furthermore, considerably more energy of both sorts is stored in the boiler than can be used at any one time. The fact that the boiler appears to do nothing except sit there and get hot should not obscure the fact that the energy it contains makes it not only the most powerful but also the most potentially dangerous part of the entire locomotive.

In this book therefore the author rightly insists upon sound methods of design and construction. Fortunately, model boilers have an excellent reputation for safety, and accidents have been few and far between. But however much experimentally-minded model locomotive builders may ignore theory and dislike figuring, in model boiler making it is a "must", unless, of course, one is following a proven design by a competent designer. The force of steam and the strength of metals make no concessions to the inexperienced or the unwary.

The efficient combustion of fuel involves raising it to a sufficiently high temperature to react with the oxygen in the air, supplying air in the right quantity, neither too little or too much, and providing space for the reaction to proceed to completion. This is the business of the ashpan, grate and firebox, and of the brick arch and combustion chamber if they are incorporated. Although the theory of combustion may be complicated, the simple rules given by the author will ensure successful results in the range of sizes he is considering.

But the mere generation of heat is of little use until it has been transmitted to the water and, by evaporation, turned into steam and the steam has been super-heated. The internal proportions of a model locomotive boiler differ markedly from those of a full-size locomotive, as those who have unsuccessfully tried to steam a "scale model" boiler will know. Thanks however to the patient experiments of many model engineers, and particularly to the late C. M. Keiller, who drew order out of chaos, the right proportions have been discovered and can be adopted with confidence.

Perhaps the biggest aid of all to model boiler makers has been the advent of highly reliable and easy to use silver solders and fluxes, which the author rightly advocates should be used wherever possible. With these and a Propane torch many of the terrors, real and imaginary, of brazing with hard spelter and a five pint blow lamp have disappeared and there is now no excuse whatever for leaving a locomotive chassis unfinished without its boiler. Steel boilers and welding, particularly in the larger sizes, are in my opinion, still best left in the hands of the expert who knows what he is doing.

The result of all this can lead to a piece of equipment of remarkable efficiency. Although not many scientifically-conducted tests of model boilers have been carried out, those by Mr. Busbridge of Cape Town University and by Mr. J. Ewins of the City University can be accepted without any doubt as to their authenticity. I know of few other examples of model engineering which achieve efficiencies as good as or better than their prototypes.

All readers of this book, therefore, owe a debt of gratitude to the many model engineers who have worked to bring this about, and to Martin Evans who has summarized the best of all their work in a single volume.

Quorn. Leics. D. H. CHADDOCK.
March, 1969.

Introduction

IT IS hardly surprising that the steam railway locomotive should still be today, long after its disappearance from the tracks of British Railways, the most popular subject for working models.

Perhaps this is partly due to the amazing variety of types and classes to be found, from the simple 0–4–0 contractor's shunting engine to the huge multi-cylinder Garratt or Mallet. I have no doubt that the extraordinary "alive" nature of the steam locomotive has a great deal to do with its fascination for enthusiasts, many of whom declare that they can tell the class of locomotive from the sound of its exhaust!

I doubt whether any other machine invented by man has proved so useful in every kind of peaceful pursuit, nor has so captured the imagination of so many people and this interest is reflected in the immense variety of books on the steam locomotive which have been published, especially during the past ten years.

When we come to the model steam locomotive however, the number of technical and descriptive books available to us is very much smaller. This is particularly so in the case of books on the locomotive boiler. Apart from the works of the late Henry Greenly, the only books that I know of on this subject are the small handbook by the late E. L. Pearce, written more than sixty years ago, and the recent work by my friend, Mr. K. N. Harris.

This is all the more surprising because it is in the design and construction of the model locomotive boiler that help and assistance are most likely to be required. There are many fine craftsmen who can and do design and build their own locomotive chassis, whether it is a simple 0–4–0 tank engine or a four cylinder Pacific express locomotive, but when they reach the boiler they seem to lose confidence and feel obliged to work to the nearest published design available.

It is also an unfortunate fact that a good many locomotive chassis are built that never see a boiler, a state of affairs which I hope this book will help in some measure to remedy. While it is true that the kind of work involved in boiler making is very different to that in chassis work and lathe work in particular, it is really no more difficult once the unfamiliar techniques have been mastered.

In this book I have tried to show that model boiler design can be safely attempted without any great knowledge or experience, provided that a few simple rules are followed. As to construction, if the inexperienced worker does not hurry the job, but tackles each section carefully and thoroughly, examining each part as it is made or assembled and rectifying any fault that may show up before proceeding to the next stage, there is no reason why he should not complete his boiler successfully. "Make haste slowly" should be the axiom in boiler work.

On the question of safety in model boilers, I have tried to stress the point that while the utmost care should be taken to ensure that any boiler built is absolutely safe, the strength factor *can* be overdone. No good purpose is served by working to a factor of safety of more than 8; in fact this is only wasteful of material and likely to lead to inefficiency in steam raising. In other words, it is bad engineering.

May I also stress that all boilers should be tested at least once a year by competent engineers and no locomotive should be steamed in public without adequate insurance, not only against accidents to the driver or his passengers, but against possible accidents to spectators.

In writing this book I have drawn freely on the work of many distinguished model engineers, some now unhappily passed on, in particular the late Henry Greenly, the late C. M. Keiller, Mr. K. N. Harris, Mr. Edgar T. Westbury and Mr. J. Ewins. I would also like to acknowledge much practical help and advice from that fine professional boiler maker, Mr. Alec Farmer.

Finally, may I record my gratitude to my distinguished friend, Professor D. H. Chaddock, C.B.E., for kindly contributing the Foreword. With his wide knowledge and practical experience in all things mechanical, no-one could have better qualifications for such a task.

Hemel Hempstead 1969. MARTIN EVANS.

CHAPTER ONE

The Modern Locomotive Boiler

BEFORE the design of the model boiler is considered, an examination of the design and construction of the full-size boiler used in the modern railway locomotive may be made with advantage.

The requirements of a locomotive boiler are very exacting, as the boiler has to withstand high steam pressures with a large margin of safety, combined with high efficiency and economy in space. The modern locomotive boiler is not a cheap one to produce, and railway engineers have made many attempts in the past to design a boiler which could be manufactured more cheaply, yet the boiler of the present day is basically very similar to that fitted to the *Rocket* of 1829.

The locomotive boiler is a quick producer of steam for its size and weight, although not so economical as the best of the stationary types. It may be described as an internally fired horizontal multitubular boiler, consisting essentially of four main parts:— The inner firebox, the outer firebox, the barrel containing the firetubes, and the smokebox. The furnace is built up in the inner firebox, and the heated gases from it make their way through the tubes or flues in the barrel, to the smokebox at the other end and thence to the chimney.

As the number of firetubes employed is large, being anything up to 250, it will be seen that the heating surface in contact with the water is very great; but even so, to generate sufficient steam it is necessary to urge the fire artificially to increase the rate of combustion. To do this the exhaust steam escaping from the engine cylinders is utilised; it is emitted from a "blast pipe" below the base of the chimney, and almost any intensity of draught can be obtained by contracting the orifice of the blast pipe. Every stroke of each piston sends a discharge of exhaust steam through the blast pipe and up the chimney, withdrawing some of the contents of the smokebox with it, and leaving a partial vacuum in the smokebox, and it is to destroy this that air rushes through the fire from beneath, causing rapid combustion.

The efficiency of a boiler is measured by the amount of water that can be evaporated per pound of fuel, and this depends on the quantity of fuel that can be consumed efficiently on the firegrate. For the economical production of steam, a boiler must not only be well designed for its work, but must be handled with intelligence.

With the constant increase in working pressures, improvements have had to be made in the methods of boiler construction. Rivet holes, which at one time were punched, are now drilled, and generally special machines are employed so that the boiler shells and plates can be drilled after rolling or pressing to shape. By this means, the holes in the plates can be made to line up with each other more exactly, and the rivets made a much better fit than was possible by the older methods.

The old-fashioned methods of riveting and caulking by hand have been superseded by the employment of hydraulic and pneumatic tools. The modern hydraulic riveter will apply a pressure of as much as 100 tons per sq. inch on the rivet head; this great force ensures the complete filling up of the rivet hole and the bringing together of the plates in a manner impossible with the old hand hammer.

In modern boiler practice, welding is being resorted to for certain parts of the boiler, having the advantage of dispensing with the drilling of holes and the fitting of rivets. It also effects a saving in weight of the finished boiler.

The boiler shell or barrel of the modern locomotive boiler is generally made up from flat mild steel plates from ½ in. to ¾ in. thick and 3 ft. and 6 ft. wide. These flat plates are rolled by "bending rolls" into circular rings of the required diameter, and joined by longitudinal seams, which are usually butted and riveted together by means of covering plates, inside and out, secured by double rows of rivets. Two, three or four of these rings are then joined together to form the barrel. The ring joints are generally of the lapped type with 4 in. to 5 in. of lap, and double rows of rivets are again used.

On some railways, notably the ex-G.W.R. and ex-L.M.S. in this country, the boiler barrel of the locomotive was of tapered shape instead of being built up from parallel rings. By adopting this arrangement, the greatest cross-sectional area is obtained nearest to the firebox where the heat value is at its greatest, while the surge of water up and down the barrel is reduced. There is also the advantage that the space in the upper part of the barrel next to the firebox is increased and it is from this part that steam is usually collected; another point is that as the distance from the rail to the top of the smokebox is reduced, the length of the chimney can be increased, giving a better draught.

The firebox wrapper or outer firebox plate is shaped to various designs, the most common being the Belpaire type, which is flat, or nearly so, on the top and sides. The round topped firebox is also commonly used, while both the Belpaire and the round topped types are used in conjunction with wide and narrow fireboxes.

Wide fireboxes are those which spread out over the top of the main or trailing frames of the engine, in order to obtain larger grate area, while narrow fireboxes are those which are waisted in towards the foundation ring so that the lower portion lies between the mainframes.

Steel plate is used for the wrapper as for the barrel, the lower edges of the wrapper being secured to the foundation ring. The wrapper is held to the

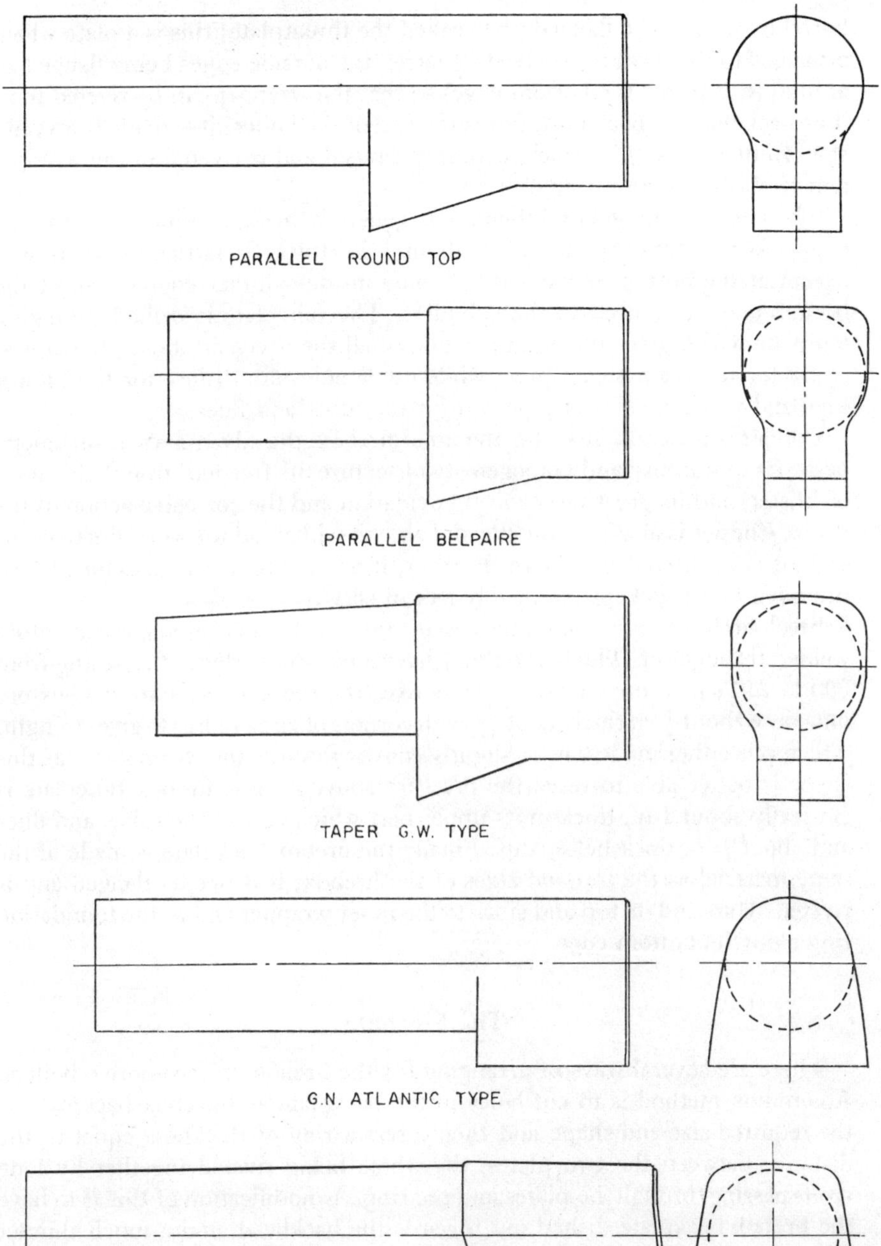

Fig. 1. Types of locomotive boilers

barrel by means of a flanged plate called the throatplate; this is a plate which is flanged to the front to receive the barrel, the outside edges being flanged all around (except for the bottom edge) so that the wrapper can be riveted to it. The backplate, or backhead, forms the end of the boiler; it is made from mild steel from ½ in. to ¾ in. thick, is deeply flanged and is riveted to the wrapper and to the foundation ring.

The front or smokebox tubeplate is generally made of mild steel, though copper is sometimes used, and is about 1 in. thick. It is flanged all round, except at the bottom in the case of some inside-cylinder engines where the bottom may be secured to the cylinders. The tubeplate is drilled through a template which gives the exact position of all the necessary holes, the largest being for the main steam pipe, while other holes are drilled for the blower pipe and vacuum exhaust pipe and for the tubes and flues.

Copper is generally used for the inner firebox, the advantages of this metal being its uniformity and homogeneity of texture, its freedom from lamination or blister, and its great resistance to oxidation and the corrosive action of the water. Copper is also very ductile and able to withstand the great fluctuations of heat encountered in service. Further, it is an excellent conductor of heat provided that it is kept reasonably free of scale.

Steel fireboxes are sometimes used, and steel inner fireboxes are often welded throughout. The firebox may have a heating surface of anything from 100 to 250 sq. ft. on a British locomotive; the top and sides are made from one piece about ½ in. thick, bent up with corners of large radius to give strength. The top is either made flat, or slightly convex towards the steam space as this shape is better able to resist the pressure above it. The firebox tubeplate is generally about 1 in. thick in its upper part which carries the tubes and flues and about ½ in. thick below this. Finally the firebox backplate is made of the same material as the top and sides of the firebox; it is deeply flanged and is riveted all around its top and sides to the inner wrapper and to the foundation ring along its bottom edge.

The Firehole

There are several ways of arranging for the firehole in locomotive boilers. A common method is to cut holes in the backplate and firebox backplate of the required size and shape and then insert a ring of thickness equal to the distance between the two plates, the whole being riveted together by long rivets passing through the plates and the ring. A modification of this is to have the firebox backplate dished out towards the backhead, and a much thinner ring riveted in; this method is often preferred to the first mentioned, as it allows greater flexibility to the firebox. A third method of making the firehole is to flange both the plates outwards so that they can be riveted directly together with no spacing ring; this entails heavier work in flanging the plates but has the advantage that the rivets are not exposed to the direct action of the fire.

Two views of a full-size locomotive boiler showing tube arrangement and staying. The boiler is from an ex-Southern Railway "Atlantic"

Neat backhead fittings on a model S.R. 4–4–0 "Schools" class locomotive to $\frac{3}{4}$ in. scale. Photograph by J. D. Sharp, Sidcup

"Full-size" backhead fittings: A view inside the cab of a L.N.E.R. 4-6-0 B.1. class locomotive

Two views to show how full-size boiler bands are arranged. Below: L.N.E.R. B.1. class

Great Western Railway 2-6-2 45xx class

Staying

Both the inner and outer firebox contain large areas of flat surface, and considerable and varying stresses are set up between them; the staying of these areas is therefore an exceedingly important matter. The circular parts of the boiler can easily withstand the internal pressure of the steam, provided they are made of a suitable thickness of metal, but the flat surface would immediately bulge unless an effective method of staying was adopted.

The inner firebox is usually tied to the outer wrapper by copper stays $\frac{7}{8}$ in. to 1 in. diameter, threaded at each end ten, eleven or twelve threads to the inch. The stays are generally pitched about 4 in. apart and are fitted through threaded holes in the inner and outer shells, these being tapped in one operation. They are screwed home with about $\frac{1}{4}$ in. projecting through both plates, the projections are then hammered down to form a head and to ensure a good steam-tight joint.

Steel stays are sometimes used in boilers, these may be $\frac{1}{2}$ in. or $\frac{5}{8}$ in. diameter, and are generally drilled at each end so that the ends can be expanded by hammering a hard steel drift into the holes. Stays made of Monel metal or Bronze have also been used. Monel metal is an ideal material for boiler stays, being highly resistant to corrosion and very strong; it is however expensive. Both Monel metal and Bronze stays may be slotted longitudinally to give greater flexibility. Some manufacturers drill a hole right through the middle of the stay, so that should one break, the steam and water will escape through it, and give warning; otherwise stay breakages can only be detected by tapping on the ends with a light hammer and carefully noting the sound produced.

The crown of the locomotive firebox was at one time supported by girder stays, but this practice is now being superseded by the use of direct stays. The most rapid evaporation in the boiler takes place on the top of the firebox, one square foot of this part of the box having been found to evaporate from six to ten times the amount of water that one square foot of average tube surface will evaporate. It is therefore essential that the circulation over the firebox crown should be as free as possible, so that the heated water can be rapidly replaced by the colder water from beneath. When girder stays are used, they pass immediately over the top of the inner firebox and therefore tend to interfere with the circulation of the water. Girder stays also make the washing out or de-scaling of the firebox more difficult.

When direct stays are fitted to the crown, the outer and inner plates are drilled and tapped and the stays screwed in from the top, a copper washer being put under the head, and nuts, fitted with copper washers, tightened up against the crown plate inside and out. Flexible stays are generally used for the first two or three rows of crownstays at the front end of the firebox, as the expansion stresses are greatest in this area.

The tubeplates are sometimes stayed by means of longitudinal rods which pass from the firebox to the smokebox, the tubeplates are also greatly strengthened by the tubes and flues, when these have been expanded in position.

Longitudinal stays are also fitted between the smokebox tubeplate and the backplate, running the full length of the boiler above the firebox; in long boilers however, these are often replaced by stays which do not run the whole length of the boiler, but have feet which are riveted to the barrel plates, or forked ends connected to angles attached to the barrel at various points in its length.

That part of the firebox tubeplate immediately below the tubes is connected to the bottom of the barrel by "palm" stays. These are special stays which are flattened at one end to take rivets for attachment to the barrel, and the other end forming a boss which is drilled and tapped to take set bolts. The set bolts, which may be of copper or steel, screwed in from inside the firebox and hammered over, or they may have square heads, with copper washers placed under the heads. In either case the tapping is done in one operation, right through the firebox plate and the boss of the stay. In some recent locomotive boilers, the palm stays have been dispensed with, by flanging the throatplate to a small radius and arranging the top row of ordinary stays as close up to the underside of the barrel as possible.

The Tubes and Flues

The tubes and flues of the locomotive boiler are made of copper, mild steel or alloy steel. The diameter of the tubes may be anything between $1\frac{5}{8}$ in. and $2\frac{1}{4}$ in., according to their length, and 13 or 12 s.w.g. in thickness. They are generally expanded into the firebox tubeplate and beaded over on the firebox side. Sometimes a steel or malleable cast iron ferrule is driven in at the firebox end to protect the end of the tube from the action of the fire. The use of the ferrule, however, has the disadvantage that the cross-sectional area of the tube is restricted somewhat at the firebox end.

When a steel firebox tubeplate is used, the tubes are usually first expanded, with about $\frac{1}{4}$ in. projecting beyond the tubeplate, and then welded to the tubeplate; at the smokebox end, the tubes are expanded. Sometimes the smokebox ends of the tubes are belled out to about $\frac{1}{8}$ in. larger diameter for a length about 3 in. before the expanding is carried out. This facilitates fitting the tubes and also their removal in the event of severe scaling or burning.

As the hot gases from the firebox pass through the tubes, some of the heat contained in them passes into the water through the walls of the tube, but a considerable portion of the heat passes right through and into the smokebox. Several means of utilising this heat have been devised, and one of the most notable is the "Serve" tube, which having longitudinal internal ribs, increases the internal heated surface of the tube, "cuts up" the gases and abstracts the heat from the centre. Good results have been obtained from these tubes, although their use means a reduction in the number of tubes as they have to be larger in diameter than normal tubes. A disadvantage of Serve tubes is the tendency for soot to lodge in the spaces between the ribs, and this is difficult to clean out.

Grates and Ashpans

The locomotive firegrate, which forms the bottom of the firebox, consists of a number of longitudinal bars, with spaces between them for the passage of the air necessary to ensure good combustion of the fuel. The grate may be entirely horizontal, or partly horizontal and partly inclined downwards towards the front end. Sometimes the whole of the grate may incline towards the front end. Provided that the incline is not made too steep, this arrangement is considered by some to be advantageous as the coal has a tendency to work down towards the front of the firebox from the firedoor. However the inclined part of the grate is generally intended by the designer to clear one of the coupled axles which may lie near to the grate at this point.

Cast iron is generally used for the firebars which are about $\frac{3}{4}$ in. thick at the top, tapering to about $\frac{3}{8}$ in. thick at the bottom, so that clinker and ashes can fall clear and into the ashpan. The depth of the bars may be between 4 in. and 5 in; they are usually made half the length of the grate, so that they can be handled more easily.

Rocking firegrates were used a great deal in America and were also used to some extent in British engines. The grate may be divided into a number of pivoted units, operated by means of levers in the cab. The rate of combustion of coal on each square foot of firegrate ranges from 40 to 120 lb. per hour, according to the type of coal and the working of the engine.

The ashpan is of box form, made from steel sheet about $\frac{5}{16}$ in. thick, one or both ends being provided with doors or dampers operated from the cab. The bottom of the ashpan is shaped so as to prevent the ashes being blown out on to the motion, and in recent designs is provided with hopper-type doors underneath so that the ashes can be easily dropped at maintenance depots and sheds. The dampers regulate the quantity of air admitted to the fire, thus the fire can be damped down by shutting all the dampers if the locomotive is required to stand by for a time.

Firedoors

Various types and shapes of firedoors have been used on locomotive boilers. At one time a plain round or oval door was used extensively, hinged at the bottom; the latch was provided with a sector having notches in it, so that the fireman could regulate the quantity of air admitted above the fire by partially opening the door on the latch. A flap plate is generally used with this type of door to protect the footplate men from the heat and to prevent the driver's eyes from being dazzled during darkness.

Another firedoor used extensively by the former L.M.S. and Great Western and other railways is the sliding type. The door is made in two halves, arranged to slide horizontally in grooves; the halves are connected together and to a hand lever; on movement of this lever, the two halves of the door move in opposite directions, to open or close the opening as required.

On some railways, engine crews were instructed to close the firedoor between each shovel of coal, it being considered that this practice led to economy in fuel and prevented damage to the firebox and tubes. However considerable effort is usually required to open and close the doors so frequently, and so the "Ajax" automatic firedoor was produced. The "Ajax" automatic firedoor can be worked by either steam or air, yet no complicated mechanism is involved. The doors, which are made in two halves, are supported on trunnions above the firehole, and a cylinder is arranged above them, fitted with a piston and valve chamber. Steam is admitted to the cylinder by means of a valve operated by a foot pedal which is depressed by the fireman as he swings his shovel towards the firehole. The piston travels forward and imparts motion to the adjacent door by means of a hardened steel roller fitted to the crosshead pin.

The two doors are geared together by quadrants, and they are lifted and kept open until the pressure on the pedal is released as the fireman turns for the next shovel of coal, when they close instantaneously. This constitutes an important safety device should a firetube burst or a "blow-back" of the fire occur.

Steam is supplied to this firedoor by a $\frac{3}{8}$ in. bore pipe connected to the steam turret on the top of the firebox; it first passes into a reducing valve, which reduces its pressue to 70 to 80 lb. per square inch. A combined control and drain valve is used to regulate the steam to the firedoor.

Brick Arches

Brick arches are still in general use in locomotive boilers. The arch is made from firebrick or fireclay lumps and is erected inside the firebox immediately underneath the bottom row of tubes. It is built in the shape of an arch so as to be self supporting and is erected from side to side of the firebox, the two ends resting on brackets or studs. Special firebricks and studs are usually inserted at intervals between the brick arch and tubeplate so that there is a small space between the face of the tubeplate and the back of the arch. The brick arch, which extends diagonally upwards towards the firehole door, acts as a baffle, deflecting the heated gases towards the back of the firebox, enabling them to mix properly with the additional air admitted above the firebed. When the arch is at a bright red heat, it assists combustion, as it helps to maintain the temperature necessary to unite the carbon of the coal and the oxygen in the air. The arch also assists in protecting the ends of the tubes from the direct action of the flames.

In addition to the brick arch, a scoop or deflector is nearly always fitted to the locomotive firedoor; this is arranged to point slightly downwards so that the cold air entering through the firedoor is directed towards the top of the fire and away from the ends of the tubes.

Combustion Chambers

In the larger types of locomotive boiler, especially those of the wide firebox type, combustion chambers are often fitted. This is an extension of the firebox into the barrel, thus increasing the volume of the firebox and its heating surface, without alteration to the grate area. At the same time, the length of the tubes and flues is reduced. The use of the combustion chamber ensures more complete burning of the fuel and enables a longer barrel to be employed than would otherwise be possible. Generally speaking, the wide round-topped or Wootten type of firebox always calls for the addition of the combustion chamber, except where the barrel is on the short side, as in some engines of the "Atlantic" type. Both the ex-L.M.S. and ex-L.N.E.R. "Pacific" locomotive boilers were fitted with combustion chambers.

Thermic Syphons

The rapid generation of steam in any boiler depends partly on the efficient circulation of the water around the firebox. In most locomotive type boilers, the natural circulation around the firebox is not all that could be desired, and to promote a more active circulation, the Nicholson Thermic Syphon is sometimes fitted.

It is claimed for the thermic syphon that in addition to improving the circulation of the water, there is a useful addition to the firebox heating surface, in a position where such a surface has its highest value. The syphon also acts as a valuable strut or support to the crown of the firebox.

The circulation in the syphon is induced in the following manner:— The water which evaporates inside the syphon, from the effect of heat in the zone of highest temperature, forms a succession of steam bubbles rising from the walls of the syphon. As these bubbles leave the walls, they accumulate in large quantities in their passage upwards through the water in the syphon. The expansion of each bubble and the accumulation of bubbles, displaces solid water, creating a constantly unbalanced head of water in the barrel of the boiler, due to its greater weight per cubic foot as compared with the much lighter combination of steam and water within the syphon. The force of gravity causes the heavier water in the barrel to rush into the syphon through the opening in the lower part of the tubeplate.

This displacement of water by steam is continuous, for as fast as the resulting current of water sweeps away steam bubbles accumulating on the walls of the syphon, other bubbles are formed. Being lighter than water, the bubbles force their way upward, pushing the water along with them, thus increasing the velocity of the water rushing in through the neck of the syphon. The volume of water kept in motion through the neck is very great and serves to continually sweep the inside walls of the syphon, washing away any sediment that might tend to lodge on them. Water entering the syphon passes through the neck at a high velocity, but as the discharge opening in the crown sheet is generally ten to twelve times the area of the neck, the water is quietly released over the crown sheet, avoiding too much turbulence at this point.

An incidental advantage of the Nicholson syphon is that if the water level in the boiler is accidentally allowed to get low, water will still continue to rise through the syphon, flooding the crown sheet and preventing overheating or collapse.

The Nicholson syphon was much used in America, where tests have shown that it will increase the evaporation of water per pound of coal by up to $7\frac{1}{2}$ per cent. The syphon has also been used on some British locomotives. Fireboxes up to 3ft. 6 in. wide are generally fitted with one syphon, fireboxes from 3 ft. 6 in. wide to 6ft. 6 in. wide, two syphons, while fireboxes over 6 ft. 6 in. wide are generally fitted with three syphons.

Construction of the thermic syphon is in steel, generally about $\frac{3}{8}$ in. thick; the bottom is bulged out and the firebricks of the brick arch rest on the shoulders so formed. The upper part of the syphon is stayed by means of direct stays, threaded through both plates and riveted over. The syphon is connected to the firebox tubeplate by means of a diaphragm or breathing plate, which is designed to absorb part of the movement of the syphon due to the expansion and contraction of the boiler. The syphon is welded to the firebox by a good ductile weld, free from pinholes. Provision for washing out and inspecting each syphon is made in the form of wash-out holes, one in the tubeplate opposite the syphon opening, one or two in the crown sheet, and one in the backplate for each syphon.

Fusible Plugs

Fusible plugs are generally fitted to locomotive fireboxes, to protect the crown sheet should the water level be accidentally allowed to fall too low. One is usually fitted to each end of the firebox, being screwed in with a tapered thread of about 11 threads per inch. The centre of the plug is bored out and countersunk at the top and the hole so-formed filled with a low melting point alloy containing lead, tin and bismuth. Sometimes the plug contains a brass or gunmetal button, sweated into the plug on the fire side with pure lead; on the water side only a very thin coating of lead is used. This type of plug generally has a longer life.

The Steam Dome

The amount of steam space above the working water level in a locomotive boiler is necessarily restricted, in order to clear the standard loading gauge. The limitation of the steam space, together with the violence of ebullition due to rapid evaporation and the motion of the engine on the track, make the question of priming and the collection of dry steam, matters of great importance. To overcome these difficulties, a steam collecting chamber or dome is generally fitted on top of the boiler barrel.

The best position for the dome is usually about the middle of the barrel, for when the engine is accelerating forwards or backwards there is a tendency for the water to rise at the ends of the boiler. The steam collecting pipe for the regulator or the regulator valve itself is placed in the dome, the inlet of the pipe being arranged at the highest possible point to assist in the collection of dry steam.

In domeless boilers, such as those of the old Great Western Railway, the collecting point for the regulator was generally in the front top corner of the firebox. Another system of steam collection was applied to some classes of locomotive on the L.N.E.R. On the top of the boiler barrel a steam collector was formed integrally with the dome; this was produced as a steel pressing riveted to the top of the boiler. Steam was admitted to the collector by a number of narrow slots in the top of the barrel, the object being to prevent water combining with the steam on the way to the regulator valve, which, together with the steam pipes, was of large size to ensure that the pressure in the boiler could be maintained all the way to the steam chests.

The Fitting of Boilers to the Frames

Locomotive boilers are normally held rigidly to the frames at the smokebox end, either by means of a saddle or by bolting extensions of the smokebox wrapper direct to the frames.

As the boiler barrel and firebox expand considerably when heated, the firebox end of the boiler cannot be held down rigidly, it must be carried by some arrangement of expansion links or brackets. The most common form of "expansion bracket" consists of an angle piece bolted or riveted to the outer firebox shell at the level of the top edge of the main frames. This angle rests on and is permitted to slide on a level bed provided for it; it is kept in place and prevented from rising by a grooved block bearing on its upper surface. The boiler is thus allowed to expand longitudinally while being held firmly in position.

Another method of holding the boiler used at one time was by means of a cross plate bolted to the frames, and fitting between the frames of the engine, slotted holes in the frames allowing longitudinal movement for the ends of the retaining bolts.

Water Feed

The usual method of delivering the feed water into a locomotive boiler is by means of internal pipes connected to the injectors via check valves on the backhead, these pipes terminating about the middle of, or in the front part of the boiler barrel, so that the water is discharged at the coolest part of the boiler. The delivery of the feed water into this part of the boiler is, however, known to bring about local and unequal expansion and contraction, with the possibility of damage to the plates and stays.

"Top-feed" has been introduced on many railways to avoid this problem.

In the top-feed arrangement designed by the late G. J. Churchward, the water is delivered through feed pipes which are carried round and up outside the boiler barrel, and then connected to check valves through which the feed water is discharged into a shoot and thence into trays, which are supported by clips fixed to the longitudinal stays. The trays are notched at the sides, and the water is thus finally delivered in the form of a number of very fine streams which drop through the steam space to the water below. In this way, the entering feed water is heated thoroughly before it is allowed to come into contact with the metallic surfaces of the boiler.

The heating of the water in the trays also brings about the deposition of much of the scale-forming impurities in the feed water; but since the trays can be readily removed, they can be kept clean and free from furry or scaly deposits more easily than is the case with the usual internal pipes.

Full-size locomotive boiler with steam-operated firedoor: S.R. "Merchant Navy" class

Below: details of a typical G.W.R. safety valve cover

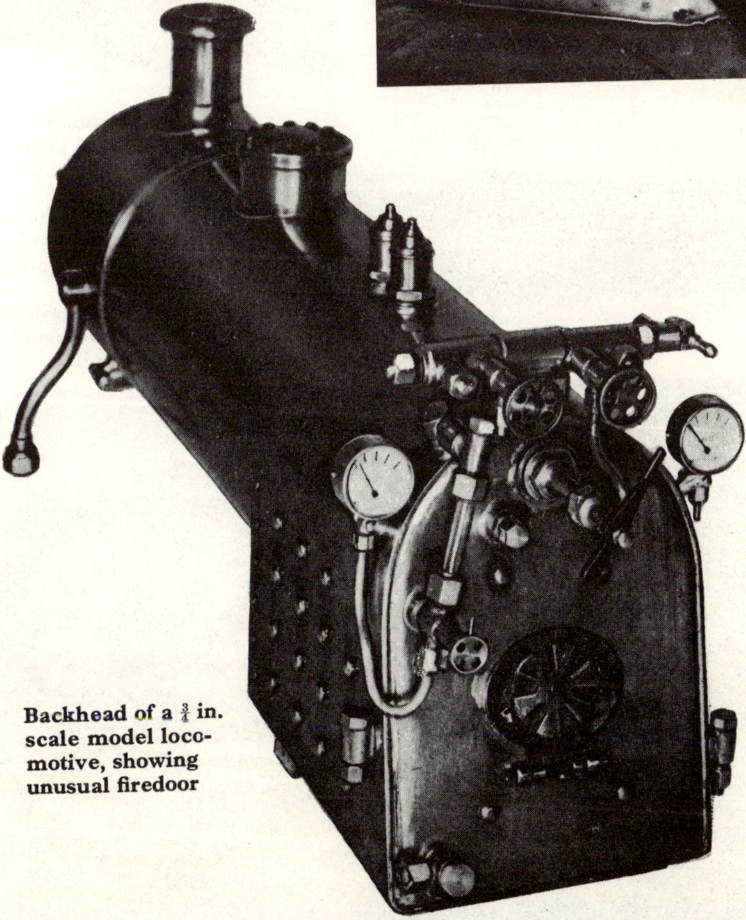

Backhead of a $\frac{3}{4}$ in. scale model locomotive, showing unusual firedoor

Crown stays on the author's "Boxhill" firebox

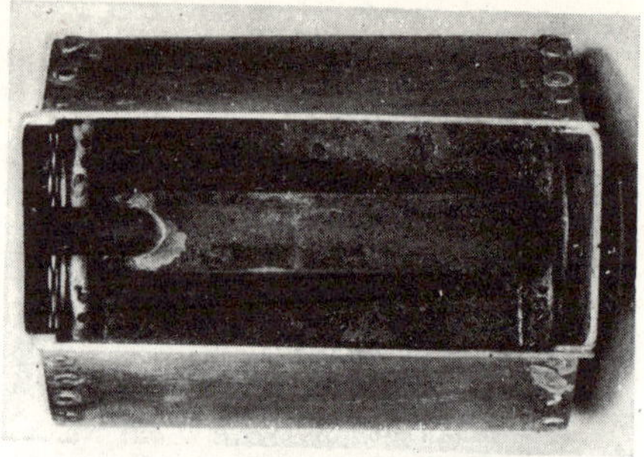

Top right: underside view of 5 in. gauge locomotive firebox showing thermic syphon.
Above: Two syphons made by Mr. D. E. Lawrence
Left: A syphon in a $\frac{3}{4}$ in. scale boiler
Below: Plate stays on tubeplate and backhead: "Boxhill" boiler

Plate 6

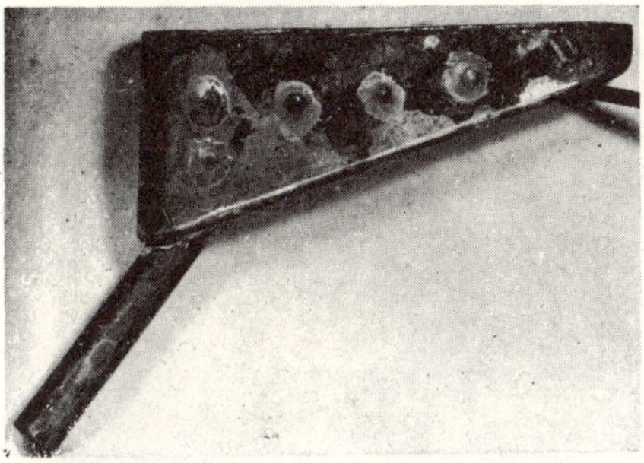

Left: a simple type of thermic syphon before fitting to boiler

Below: A fine "Britannia" boiler built by A. Palmer of South Africa

Bottom: Neat backhead fittings on a $\frac{3}{4}$ in. scale G.W.R. 4–6–0. Photograph by G. Cashmore, built by Ardis Marchant

Right: A well-filled backhead on Mr. R. Curtis's locomotive "Abertaive"

Below: Barrel made from sheet material with silver-soldered joint

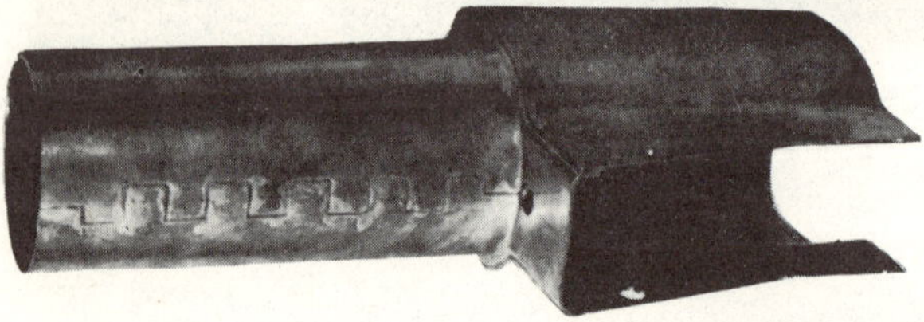

Below: Tubes placed in tubleplate ready for silver-soldering.

Below: A fine "Britannia" boiler built by A. F. Farmer

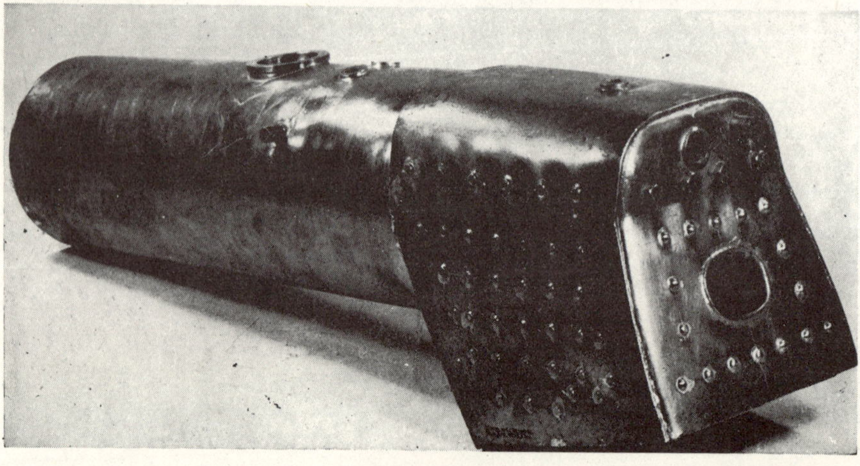

CHAPTER TWO

Designing Boilers for Models

THE great majority of model steam locomotives built today are made more or less to scale, so that the overall size of the model boiler is immediately limited to some extent by the loading gauge. This is perhaps as well, for no-one wishes to see a return to the practice of placing an outsize boiler upon a chassis of moderate proportions, nearly always with disastrous results.

The builder of the free-lance model can of course please himself as regards the overall proportion of both his boiler and chassis, subject to the limitations of the rail gauge selected. He will however be well advised to design the boiler of the model in proportion to the total cylinder capacity of the engine.

While it is true that the famous locomotive engineer of the old Great Northern Railway, Mr. H. A. Ivatt, claimed that the power of a locomotive was measured by the capacity of its boiler to boil water, this is really only a half-truth. The fact is that the proper functioning of the locomotive boiler depends on, among many other things, the correct amount of air being drawn through its firebox and tubes as a result of the action of the exhaust steam being discharged through the smokebox and chimney, and the exhaust of course depends on the cylinders, valve gear and exhaust arrangements generally.

When building a scale, or near to scale, model of a full-size locomotive which has proved successful in service, the author's opinion, borne out by many practical examples, is that the exterior proportions of the prototype boiler can be reduced almost exactly to scale, though the inside arrangements will of course have to be altered drastically. It is however advisable to increase the firebox volume slightly in most cases, without increasing the length of the barrel. One reason for this is that one square inch of firebox heating surface is worth a great deal more than the same area of tube heating surface; another reason may be stated as follows:— In dealing with locomotives having crank axles, valve gear eccentrics, or both, on the axle immediately in front of the firebox, it will always be found that the throatplate has to be set a little further to the rear of the engine than the exact scale equivalent, owing to the generally over-scale proportions of the components. This loss in firebox length should therefore be compensated for, by extending the firebox further into the cab. The increase in tube length can be allowed for by setting the smokebox tubeplate further to the rear: the question of tube proportions will however be examined later.

Having now determined the overall dimensions of the boiler, the next matter to be decided is at what height to put the crown of the firebox.

It will be found convenient for the purpose of comparison to dimension the height of the crown of the firebox from the longitudinal centre-line of the barrel. As the steam space of a Belpaire firebox is greater than that of a round-top firebox of similar length, it follows that the firebox crown can be placed a little higher in the Belpaire type. When a boiler having a tapered barrel and Belpaire firebox is considered (such as fitted to most ex-G.W.R. locomotives, and ex-L.M.S. Stanier engines), both the steam and the water spaces are relatively greater at the firebox end than in the corresponding boiler with parallel barrel, thus the crown of the firebox may again be placed a little higher.

For model locomotive boilers from $\frac{1}{2}$ in. to 3 in. scale, a good average height for the firebox crown is $\frac{7}{64}$ in. above the longitudinal centre-line of the barrel for each inch of barrel diameter, this figure being for boilers of the parallel barrel-round top firebox type. For boilers with parallel barrels and Belpaire fireboxes, this figure can be increased to $\frac{1}{8}$ in. for each inch of barrel diameter, while for taper boilers with Belpaire fireboxes, $\frac{5}{32}$ in. per inch of barrel (the barrel being measured at the firebox end) can be used.

If these figures are adopted, the following approximate heights of firebox crown above barrel centre-line are arrived at:

3 in. dia. parallel barrel, round-top	$\frac{11}{32}$ in.
3 in. dia. parallel barrel, Belpaire	$\frac{3}{8}$ in.
3–3$\frac{1}{4}$ in. dia. taper barrel, Belpaire	$\frac{1}{2}$ in.
4 in. dia. parallel, round-top	$\frac{7}{16}$ in.
4 in. dia. parallel, Belpaire	$\frac{1}{2}$ in.
4–4$\frac{1}{4}$ in. dia. taper, Belpaire	$\frac{21}{32}$ in.
5 in. dia. parallel, round-top	$\frac{9}{16}$ in.
5 in. dia. parallel, Belpaire	$\frac{5}{8}$ in.

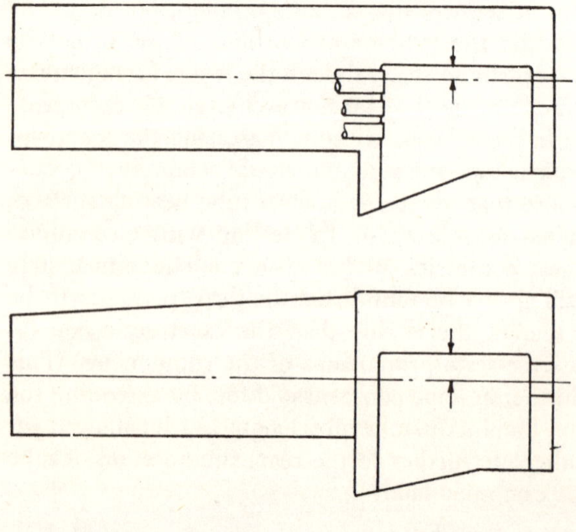

Fig. 2.
Showing the recommended height of the firebox crown above barrel centre-line

5–5½ in dia. taper, Belpaire	$\frac{7}{8}$ in.
6 in. dia. parallel, round-top	$\frac{21}{32}$ in.
6 in. dia. parallel, Belpaire	$\frac{3}{4}$ in.
6–6½ in. dia. taper, Belpaire	1 in.

The Firebox

The shape of the outer firebox wrapper can now be considered. The four basic shapes of firebox are the round-top narrow type, the Belpaire narrow type, the round-top wide type, or Wootten firebox, and the Belpaire wide type. By narrow type firebox is meant one which is sufficiently narrow across its lower part to lie between the main frames of the locomotive, while the wide type of firebox spreads out across nearly the full width of the engine, either above trailing frames or above small diameter coupled wheels as in some goods engines.

When deciding on the overall width of the part of a narrow type firebox which lies between the frames, allowance must be made for the thickness of the stay heads on each side, and also a small additional allowance for the expansion of the firebox when hot. For ½ in. scale locomotives about $\frac{1}{16}$ in. should be allowed on each side, $\frac{3}{32}$ in. for ¾ in. scale fireboxes, about $\frac{7}{64}$ in. for one inch scale, and ⅛ in. for 1½ in. scale.

The foundation ring is generally made from square or rectangular bar, and arranged so that there is an equal width of water space all around the inner firebox. Where the throatplate and firebox tubeplate lie vertical (parallel to one another), it is advisable to increase the width of the front section of the foundation ring—to about 25 per cent more than the width of the remainder—in order to assist water circulation, which is never very good in this part of the boiler. If the boiler is to be an all-silver soldered or brazed one, a very heavy foundation ring should be avoided or difficulty may be experienced in raising the temperature of this part of the boiler for successful brazing; in this case, the use of rectangular bar is to be preferred.

When deciding upon the shape of the inner firebox, it should be remembered that a plate curved so as to present its convex side to the pressure of the steam is always stronger than a flat plate, in addition a curved plate has a slightly larger heating surface. The crown of the inner firebox should therefore be made curved wherever possible. The shape of the firebox in end elevation will generally follow the contour of the outer wrapper, except that the width of the water space should increase gradually all the way up from the foundation ring in order to assist circulation.

The position and size of the firehole are important considerations in a model locomotive boiler. If made to scale, the firehole would be much too small for easy firing, especially on the run. Even in a half inch scale boiler, the firehole should not be made less than 1½ in. diameter, or to an oval shape of similar area; althouth a little heating surface may be lost through the use of such a large firehole, this is not enough to have any appreciable effect.

The firehole in the model boiler should be placed as high as possible in the firebox backplate, so as to enable the driver to observe the condition of the fire without having to bend too low; this applies even more to locomotives worked on ground level tracks than on elevated tracks.

Strength of Model Boilers

The strength of a model steam boiler is of course a most important consideration. While it is quite easy to arrive at the theoretical strength of the barrel of the boiler, the other parts of a locomotive type boiler need careful examination.

For a given thickness and diameter, a length of seamless solid-drawn tube will make the strongest barrel. When the barrel is made from flat sheet rolled or bent to a cylindrical form with the joint lapped and riveted or brazed, the strength depends on the type of joint used. For double riveting it is customary to take a figure equal to 70 per cent of the strength of a solid-drawn tube, for a single riveted joint 50 per cent is usually taken, while for a brazed or silver soldered joint 80 per cent is generally accepted.

To arrive at the right thickness of metal to be used for a boiler barrel, the following simple formula can be recommended:

$$P = \frac{D \times F \times WP}{S \times R \times C \times T \times 2}$$

where P = Plate thickness in inches.
D = Outside diameter of barrel in inches.
F = Factor of safety.
WP = Working pressure in lb. per sq. in.
S = Ultimate tensile strength of the material in lb. per sq. in.
R = Riveting allowance.
C = Corrosion allowance.
T = Temperature allowance.

The factor of safety has been taken at between 6 and 10, but 8 is suggested as quite sufficient for the purpose.

The ultimate tensile strength of the material varies a great deal according to its composition, but the following figures can be recommended:

Mild steel	60,000 lb. per sq. in.
Copper	25,000 ,, ,, ,, ,,
Stainless Steel	70,000 ,, ,, ,, ,,

Riveting allowance:

0·5 is recommended for single riveting.

0·7 is recommended for double riveting.

0·8 is recommended for welded, brazed, or silver soldered joints.

Corrosion allowance:

No allowance need be made for copper or stainless steel, but for mild steel boilers, 0·5 is recommended for plate thicknesses under $\frac{1}{4}$ in., 0·75 for plate $\frac{1}{4}$ in. thick or over.

Temperature allowance:

Copper has diminished strength at high temperatures, and therefore a temperature allowance should be made for this metal, depending on the working pressure. For working pressures of 60–100 lb. per sq. in., a temperature allowance of 0·80 may be made, and for pressures from 110–150, an allowance of 0·70 is recommended. No temperature allowance need be made for mild or rustless steels.

To take two examples of the working of this formula, first a $\frac{3}{4}$ in. scale copper barrel 4 in. outside diameter may be considered, with silver-soldered seams, together with a working pressure of 90 lb. per sq. in. Then:

$$\text{Thickness of material} = \frac{4 \times 8 \times 90}{25{,}000 \times 0{\cdot}8 \times 0{\cdot}8 \times 2} \text{ in.}$$

$$= 0{\cdot}090 \text{ in.}$$

In practice, $\frac{3}{32}$ in., or 13 s.w.g. metal would therefore be used in this example.

Sometimes the situation arises where a boiler barrel of a definite thickness is available, and it is desired to know the highest pressure at which the boiler can be safely worked. The formula can then be re-arranged as follows:

$$\text{Safe working pressure} = \frac{P \times S \times R \times C \times T \times 2}{D \times F} \text{ lb. per sq. in.}$$

To take an example, suppose the barrel is in mild steel with double-riveted joints, and is 6 in. outside diameter, with a thickness of $\frac{1}{8}$ in.

$$\text{Then working pressure} = \frac{\frac{1}{8} \times 60{,}000 \times \frac{7}{10} \times \frac{5}{10} \times 2}{6 \times 8}$$

$$= 109 \text{ lb. per sq. in. (approx)}$$

This formula should not be used for deciding the thickness of the firetubes or the superheater flues of the model locomotive boiler, as these must be made thicker than would be required on the grounds of safety alone, in order to withstand the constant abrasive action of the products of combustion; in addition the tubes and flues also act as longitudinal stays.

There is a further point to note here:— a firetube or superheater flue must also be made thicker than given by the above formula because an externally loaded tube will collapse by buckling at a lower load than that at which it will burst if subject to internal pressure.

By the formula, a $\frac{3}{8}$ in. dia. seamless copper boiler tube in a boiler working at 80 lb. per sq. in. would need to be only 0·008 in. thick,—far too thin for the purpose. Similarly, a 1 in. dia. copper flue tube in an 80 lb. boiler would, by the formula, work out at only 0·016 in. thick; again much too thin.

The following table therefore gives suitable thicknesses of seamless tubes and flues for boilers working at 80 to 120 lb. per sq. in.

TABLE I

Outside dia.	*Thickness: Copper tube*	*Thickness: Mild steel tube*	*Thickness: Stainless steel*
$\frac{1}{4}$ in.	24 s.w.g.	—	26 s.w.g.
$\frac{5}{16}$ in.	22 s.w.g.	—	24 s.w.g.
$\frac{3}{8}$ in.	20 s.w.g.	—	24 s.w.g.
$\frac{7}{16}$ in.	20 s.w.g.	—	22 s.w.g.
$\frac{1}{2}$ in.	20 s.w.g.	16 s.w.g. (or $\frac{1}{16}$ in.)	20 s.w.g.
$\frac{5}{8}$ in.	20 s.w.g.	16 s.w.g. ,,	20 s.w.g.
$\frac{3}{4}$ in.	18 s.w.g.	16 s.w.g. ,,	20 s.w.g.
$\frac{7}{8}$ in.	18 s.w.g.	16 s.w.g. ,,	18 s.w.g.
1 in.	18 s.w.g.	16 s.w.g. ,,	18 s.w.g.
$1\frac{1}{8}$ in.	18 s.w.g.	16 s.w.g. ,,	18 s.w.g.
$1\frac{1}{4}$ in.	16 s.w.g. (or $\frac{1}{16}$ in.)	16 s.w.g. ,,	18 s.w.g.
$1\frac{3}{8}$ in.	16 s.w.g. ,,	16 s.w.g. ,,	18 s.w.g.
$1\frac{1}{2}$ in.	16 s.w.g. ,,	13 s.w.g.	18 s.w.g.

Strength of Boiler Plates

We can now consider the strength of the parts of the boiler other than the barrel. The outer firebox wrapper, when made separately from the barrel, should never be made of material thinner than the barrel. In fact, if the wrapper is of the Belpaire type, it may be made very slightly thicker with advantage, as this reduces the problem of staying without adversely affecting steaming.

The inner firebox wrapper should, however, not be made unnecessarily thick, or the steam raising capacity of the boiler would suffer to some extent. If the inner wrapper is of copper or stainless steel, it can be made two-thirds to three-quarters the thickness of the barrel material. This rule cannot be applied to mild steel boilers, owing to the rust problem. It is not advisable to use plates less than $\frac{1}{8}$ in. thick in any part of a steel boiler.

The smokebox tubeplate, being a flat plate, should be made somewhat thicker than the barrel material as it generally has to be tapped to take the longitudinal stays, main steam pipe, etc. The firebox tubeplate and the firebox backplate may be made of the same thickness as the barrel, though if the boiler is a large one, and the tubes are to be expanded into the firebox tubeplate, this plate should be made 20 to 25 per cent thicker.

The thickness of the backhead of the model locomotive boiler depends to some extent on whether it is intended to screw the fittings such as the water gauge, blower valve and so forth, directly into it. In the smaller scales, where copper plates are almost universally adopted, it is a good policy to use bushes or pads for all fittings, so that the threads of the fittings have a firm and steam-tight hold in the boiler. In this case, the thickness of the backhead need not be more than that of the smokebox tubeplate. In large model steel boilers, the backhead may be made 15–25 per cent thicker than the smokebox tubeplate, or 30–50 per cent thicker than the material used for the barrel itself.

The Tubes and Flues

We have already dealt with the thickness or gauge of tubes and flues to be used in the model boiler; there is now the question of their diameter in relation to their length to be considered, and also their spacing or pitch in the tubeplates.

There is a definite relation between the diameter and the length of a boiler tube, whatever the size of the boiler. As the size of the tube is varied, the amount of gas in it—that is its volume—varies as the cube, while its surface, which has to extract the heat from the gases, varies as the square of its dimensions; so to keep a constant ratio between volume and surface, any variation in the length should be followed by an equal variation in the square of the diameter.

An examination of the most successful full-size locomotive boilers shows that the length of the tubes divided by the square of the internal diameter works out at between 50 and 70. For model boilers of all scales therefore, the following formula is recommended:

$$d = \sqrt{\frac{L}{65}}$$

where d is the internal diameter of the tube, and L is the distance between tubeplates, both being measured in inches.

This formula cannot of course be used exactly, as it is generally only possible to obtain tubes to the nearest $\frac{1}{16}$ in. diameter.

To take some examples of the working of this formula, suppose that the distance between the tubeplates of the boiler under consideration is 7 in. Then

$$d = \sqrt{\tfrac{7}{65}} \text{ in.}$$
$$= 0{\cdot}33 \text{ in. approx.}$$

The nearest standard size tube to meet this requirement would therefore be $\frac{3}{8}$ in. O.D. × 20 s.w.g. wall thickness.

Again, if the distance between tubeplates is 11 in. Then:

$$d = \sqrt{\tfrac{11}{65}} \text{ in.}$$
$$= 0{\cdot}41 \text{ in.}$$

The nearest standard size tube in this example would be $\frac{1}{2}$ in. O.D. × 20 s.w.g.

Finally, for a boiler with a distance between tubeplates of 20 in.—

$$d = \sqrt{\tfrac{20}{65}} \text{ in.}$$
$$= 0{\cdot}56 \text{ in. approx.}$$

The nearest standard size tube in this example would be $\frac{5}{8}$ in. O.D. $\times$ 20 s.w.g.

The Superheater Elements

The size of the superheater elements is governed by two considerations: the diameter must be small enough to ensure that all the steam passing through the element is properly heated, and there must be sufficient area through the elements to prevent wire-drawing. Reference to successful full-size locomotive boilers is again very helpful. Generally speaking, the shorter barrel boilers give the least trouble as far as superheat temperature is concerned. In most cases, the fraction L/2 lies between 230 and 280, where L is the overall length of the element, and d is the external diameter of the element, For model work of any scale, the formula

$$d = \sqrt{\frac{L}{275}} \text{ may be safely used.}$$

For all practical purposes, the overall length of the element may be taken as twice the distance between tubeplates. To take some examples: For a boiler in which the distance between tubeplates is 8 in.,

$$\text{outside diameter of superheater elements} = \sqrt{\tfrac{16}{275}} \text{ in.}$$
$$= 0{\cdot}24 \text{ in. approx.}$$

In this case, elements $\frac{1}{4}$ in. outside diameter would be satisfactory. For a boiler having a distance between tubeplates of 20 in.,

$$\text{outside diameter of elements} = \sqrt{\tfrac{40}{275}} \text{ in.}$$
$$= 0{\cdot}38 \text{ in. approx.}$$

Here the elements could be made $\frac{3}{8}$ in. outside diameter.

To find a suitable size of superheater flue for the elements, it is only necessary to add sufficient clearance for the passage of the gases around the elements. The following table gives recommended inside diameters of flues:—

Sizes of Superheater Flues

element, outside diameter	*flue, inside diameter*
$\frac{1}{8}$ in.	$\frac{11}{32}$ in.
$\frac{5}{32}$ in.	$\frac{7}{16}$ in.
$\frac{3}{16}$ in.	$\frac{1}{2}$ in.
$\frac{7}{32}$ in.	$\frac{9}{16}$ in.
$\frac{1}{4}$ in.	$\frac{11}{16}$ in.
$\frac{9}{32}$ in.	$\frac{3}{4}$ in.
$\frac{5}{16}$ in.	$\frac{27}{32}$ in.
$\frac{3}{8}$ in.	1 in.
$\frac{7}{16}$ in.	$1\frac{1}{8}$ in.
$\frac{1}{2}$ in.	$1\frac{1}{4}$ in.

Plate 8

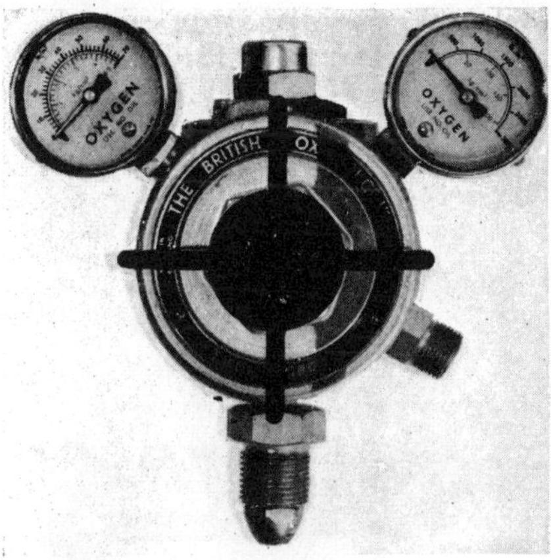

Oxygen gauges and regulators by B.O.C. Ltd

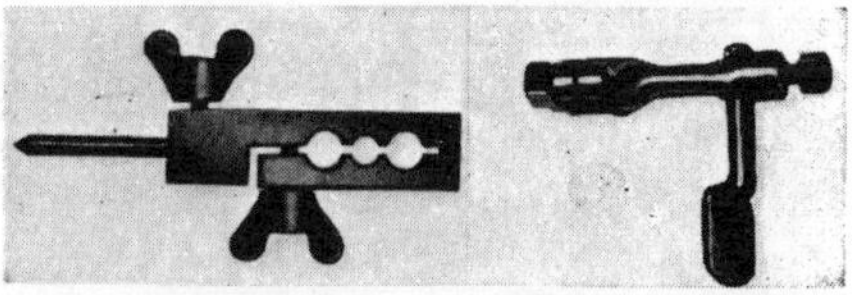

Top right: Goggles suitable for welding and brazing

Below right: Guides for flame-cutting, Acetylene gauges and regulators and "Alda" h.p. gas tubing by B.O.C. Ltd

Below: Nozzle cleaners for blow pipes by B.O.C. Ltd

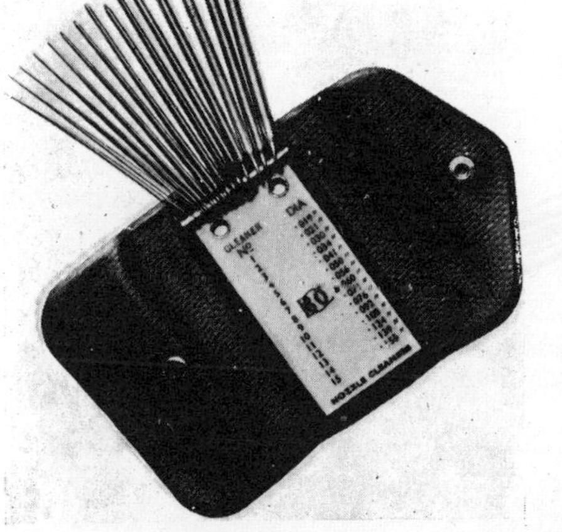

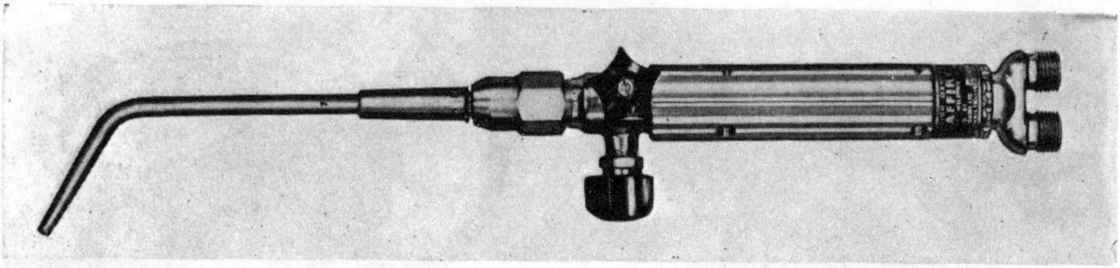

The "Saffire" lightweight blowpipe for oxy-acetylene

Below: "Saffire" with circle-cutting attachment

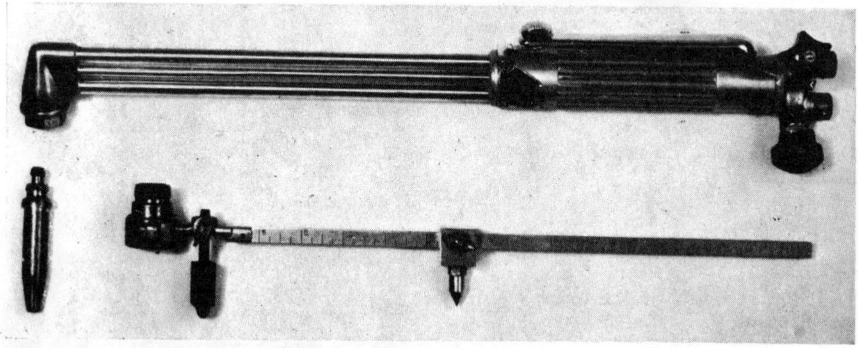

Below: "Firefly" combined blowpipe by B.O.C. Ltd

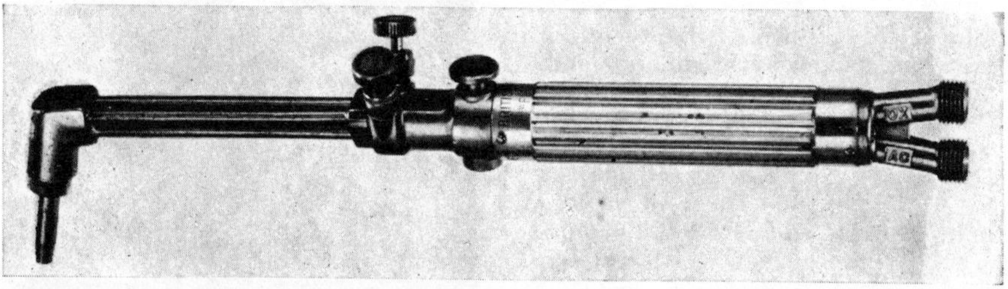

Below: "Saffire" universal cutter by B.O.C. Ltd

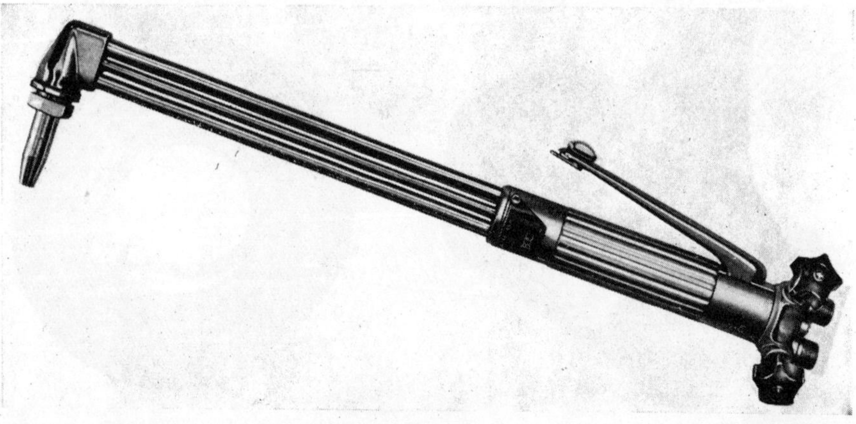

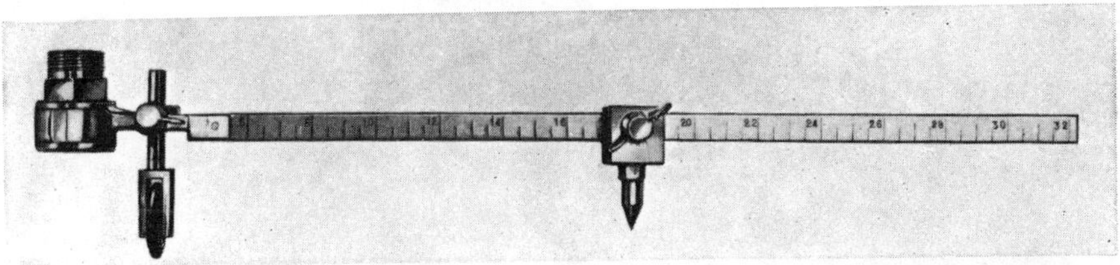

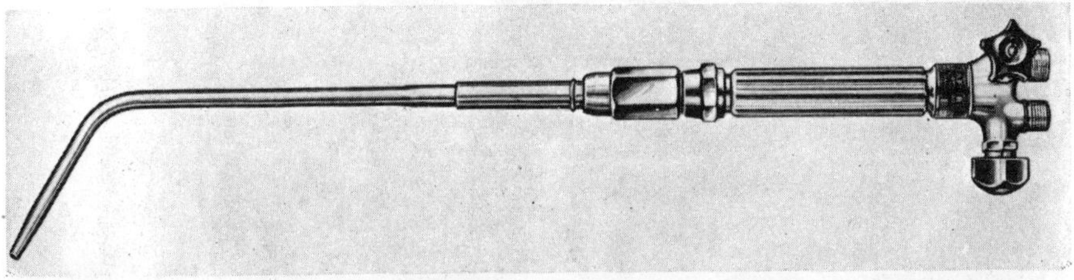

Above: Circle cutting attachment and oxy-acetylene "Saffire" blowpipe by B.O.C. Ltd

Right: Spark-lighter by B.O.C. Ltd

Below left: Double roller guide and

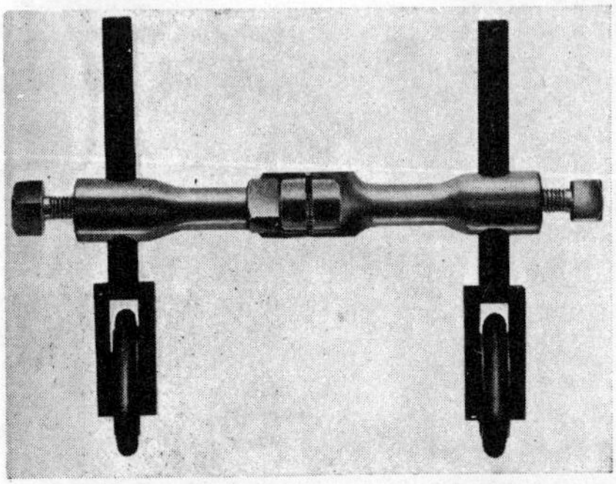

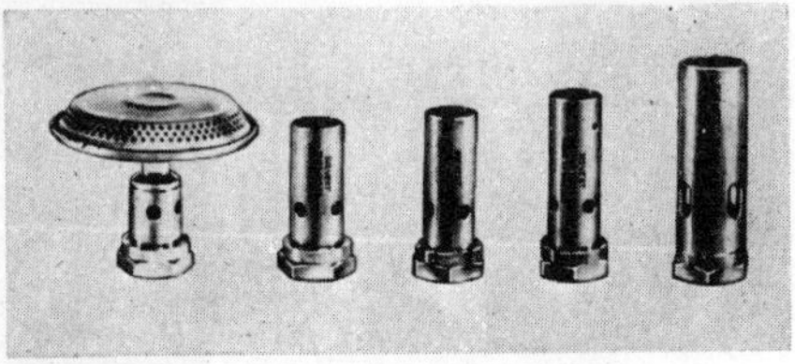

some of the Sievert burners for propane

Below: A hose failure valve for propane: Photograph courtesy W.A. Meyer Ltd

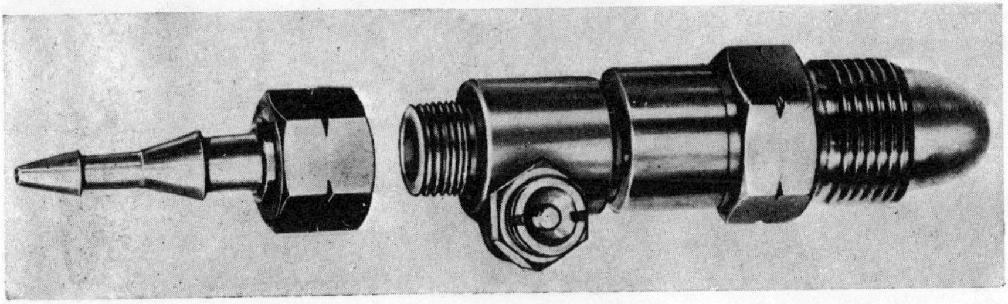

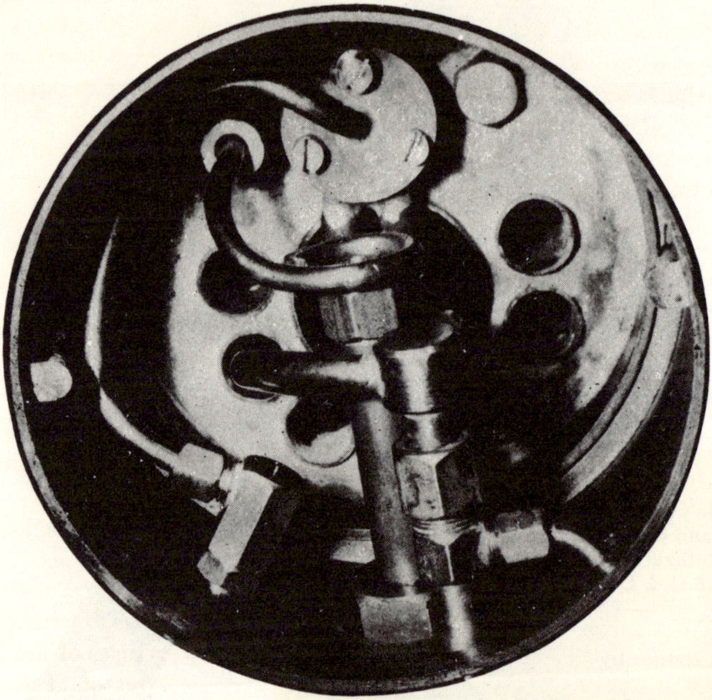

Inside of model locomotive smokebox, showing blastpipe and steam connections, also snifter valve.

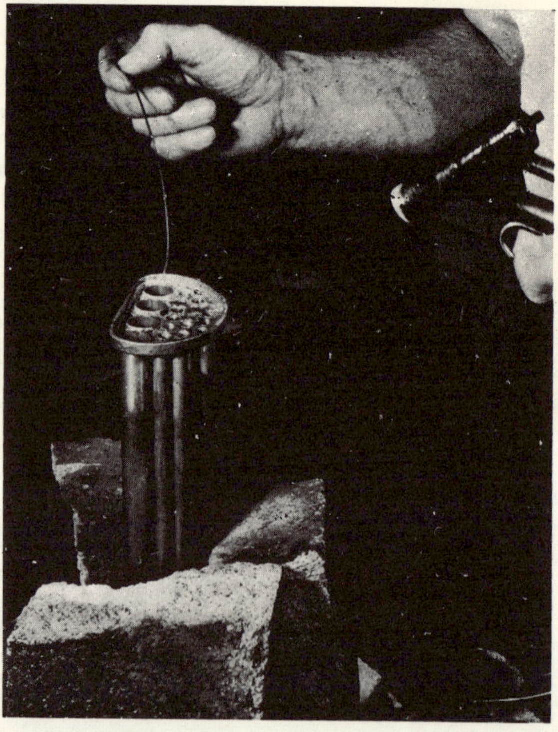

Silver soldering tubes to combustion chamber tubeplate, using an air-gas blowpipe.

REFLECTIONS ON NORTH EAST STEAM.

TRANSPORT VIDEO PUBLISHING.

P.O. BOX 229
WHEATHAMPSTEAD,
HERTS.
AL4. 8SD.

£17-95.

The total cross-sectional area of the elements must be sufficient to prevent any possibility of wire-drawing. The main steam pipe should always be made equal to at least one-quarter of the cylinder bore (for a two-cylinder engine), preferably slightly larger than this. To ensure that the superheater does its job properly, the total cross-sectional area of the superheater elements should be about 25 per cent greater than that of the main steam pipe. If the cylinders of the engine are of gunmetal, with soft packings, this proportion might well lead to steam temperatures that are too high, in which case the superheater elements could be kept to 10–15 per cent greater.

Finally we come to the actual spacing of the tubes and flues in the tube-plates. There are three ways in which the tubes can be arranged:— as a horizontal diamond pattern, in a vertical diamond pattern, or in a square pattern. The horizontal diamond arrangement is most used in models because it enables the maximum number of tubes to be used in the rather limited space available. The circulation of the water is slightly better in the vertical diamond arrangement, and better still in the square arrangement. This last arrangement does not permit the use of quite so many tubes but it can be recommended in large scale boilers.

The superheater flues should wherever possible be placed in the top row, above the ordinary tubes. There are three reasons for this. It allows for easier tube sweeping; the possibility of the flue being blocked by particles of ash, etc., is reduced; the temperature of the gases is higher towards the top of the tubeplate.

Fig. 3.
Right: Horizontal diamond tube arrangement

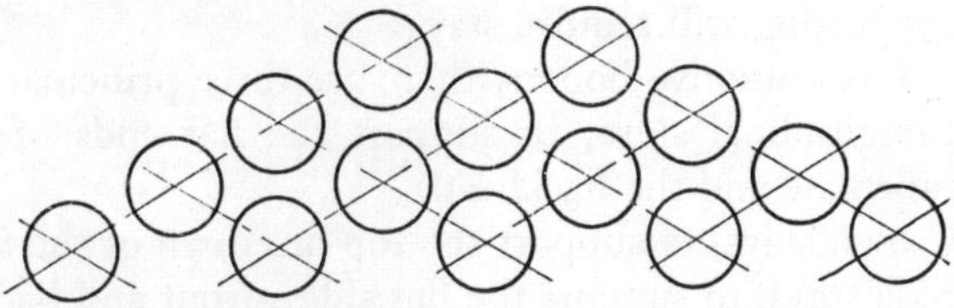

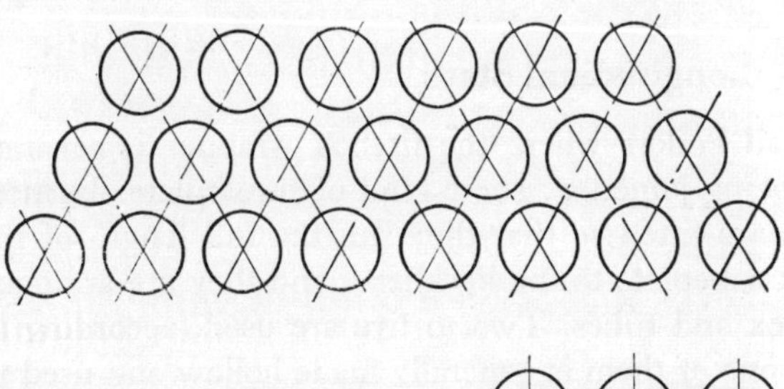

Left: Vertical diamond tube arrangement

Right: Square or Marine tube arrangement

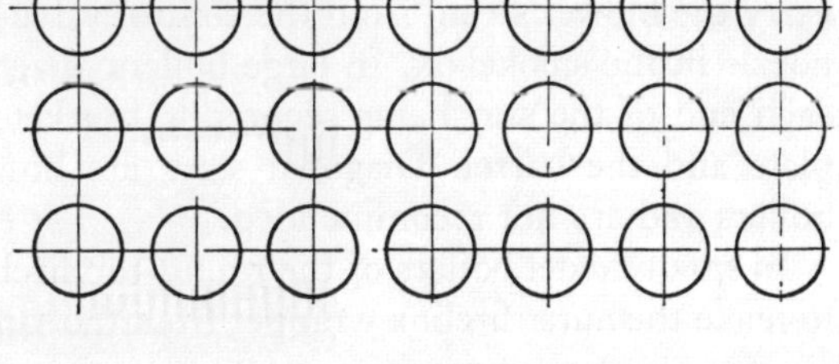

Probably the best way to lay out a tube arrangement is to mark out the available space on a drawing of the firebox tubeplate, showing the inside diameter of the boiler barrel to allow for the necessary clearances, and then to draw in the required number and size of superheater flues as decided beforehand, along a horizontal line placed high on the tubeplate. The ordinary tubes are then drawn in, and if the more usual horizontal diamond arrangement is to be adopted, the top row can be marked out on a horizontal line immediately below the flues and with a pitch spacing of Diameter $\times \frac{4}{3}$, the centre-lines of each subsequent row of tubes being Diameter $\times \frac{9}{8}$ below one another.

A closer spacing of tubes than this is not recommended, or the circulation of the water—never very good in locomotive type boilers—will be impaired. Care must also be taken to ensure that the lowest tubes do not lie too close to the underside of the boiler barrel.

Sometimes all the tubes and flues in a locomotive boiler may be given a slight rise from the firebox tubeplate to the smokebox tubeplate. This idea can be recommended as it makes the operation of tube sweeping an easier one: it may also assist the draught very slightly.

Staying

The general principle underlying the staying of a boiler is that all flat surfaces in the boiler must be supported, and in fact all surfaces not self-supporting will require stays.

In locomotive boilers, there are three principal classes of stay:

Longitudinal stays, to support the flat ends of the boiler (the smokebox tubeplate and the backhead).

Crown stays, to support the top or crown of the firebox.

Side stays, to support the flat sides, front and back of the firebox.

Longitudinal Stays

These are essential in all boilers where the firebox wrapper is separate from the barrel, the two being joined by some kind of throatplate. In most model boilers, these stays are fitted so that they run the full length of the boiler, from the smokebox tubeplate to the backhead, and they are arranged just above the inner firebox and tubes. Two to five are used, according to the size of the boiler, and one of them is generally made hollow and used to carry the blower steam from the control valve on the backhead to the blower nozzle in the smokebox. In large boilers, diagonal stays are sometimes used, each end of the stay being secured to brackets or angles riveted to the tubeplate and the barrel. Diagonal stays are however difficult to fit in model boilers and are not recommended.

In small model boilers of the round-top firebox type, it is often convenient to make the outer firebox wrapper from the same material as the barrel. Thus

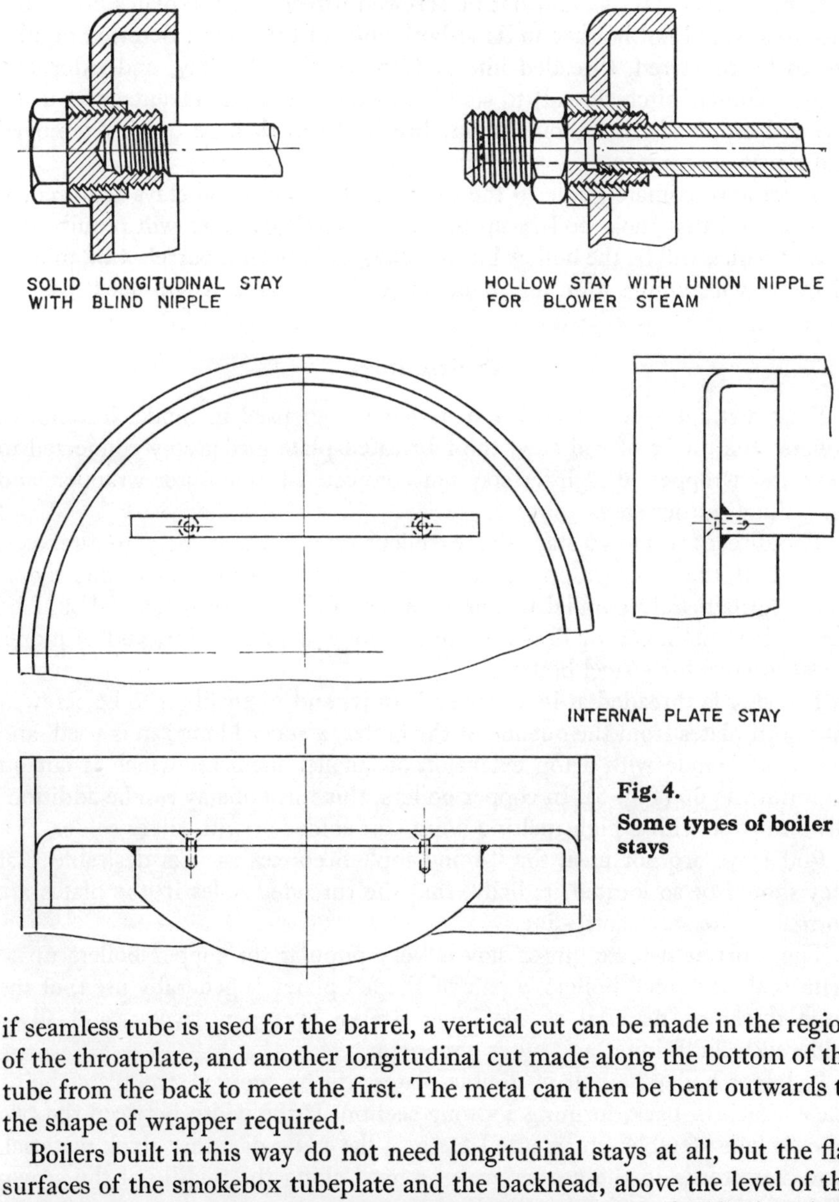

Fig. 4.
Some types of boiler stays

if seamless tube is used for the barrel, a vertical cut can be made in the region of the throatplate, and another longitudinal cut made along the bottom of the tube from the back to meet the first. The metal can then be bent outwards to the shape of wrapper required.

Boilers built in this way do not need longitudinal stays at all, but the flat surfaces of the smokebox tubeplate and the backhead, above the level of the tubes and inner firebox will need support. A simple way of stiffening these parts, used successfully by the author, is to fit a curved plate on the inside, by screwing and brazing (Fig. 4).

Longitudinal stays may be made in monel metal, drawn gunmetal or copper, in that order. Steel may however be used in steel boilers for very large models, but steel stays should never be used in copper boilers, owing to differential expansion problems and possible galvanic action.

Longitudinal stays should not be screwed direct into the plates, as in any case this would not be easy to do at both ends of the boiler. Screwed nipples are to be preferred, threaded internally to receive the stay, and externally (with a similar pitch thread) to screw into the plate. An advantage of using screwed nipples is that should a stay break, it can then be quickly removed and a new stay put in.

As an approximate guide to the number of longitudinal stays required, it may be said that most boilers up to 3½ in. barrel diameter will require two (one of which will be the hollow blower stay), boilers with barrels 4 in. to 8 in. diameter will require four, and boilers over 8 in. diameter will need five.

Crown Stays

Four distinct types of crown stays have been used in model locomotive boilers: the direct or rod stay, the fabricated-plate girder stay connected to the outer wrapper, the girder stay not connected to the outer wrapper, and the "sling" girder stay.

The direct rod crown stay is an excellent type to use in Belpaire fireboxes where both the outer wrapper and the crown are flat, or nearly so, and therefore approximately parallel to one another. This variety of stay should be made of monel metal or drawn gunmetal for a copper boiler, and of monel metal or steel for a steel boiler.

The stay is threaded at its outer ends only, and to enable it to be screwed into both plates from the outside of the boiler, a special long tap is used, and the stay is made with a top extension of smaller diameter which is cut off after nutting up (Fig. 5). In copper boilers, this form of stay can be additionally run over with a high-melting point soft solder or with silver solder.

Rod stays are not ideal for "round-top" fireboxes as it is desirable that they should be so located, radially, that the threaded holes in the plates are normal to the stay centre-line.

The fabricated-plate girder stay is very popular for copper boilers up to 1½in. scale. In small boilers, a pair of shaped plates is generally used, of the same thickness material as the inner firebox wrapper, riveted and silver soldered to both inner and outer wrappers.

In larger boilers, a pair of double plates will be adopted, the plates being placed back-to-back, forming a strong section. If the width between the two pairs is considerable, an inverted tee or a flat plate of rather thick material, bowed upwards in the centre, may be used along the middle of the crown of the firebox for its full length (Fig. 6).

When designing plate crownstays, it is most important to carry them across the full length of the inner firebox assembly, so that they lie over the flanged edges of the firebox tubeplate and the firebox backplate.

In comparing these first two types of crown stays, it must be remembered that the circulation of the water around the crown of the firebox is greatly impeded by the normal plate girder stay, and this lack of circulation leads to slower steam raising and quicker accumulation of scale. The author has therefore evolved a plate stay in which slots are cut through the feet of the stay in order to allow free circulation of the water (Fig. 6). When fitting this type of stay to the inner firebox, it is generally sufficient to use rivets at the extreme ends only, and to rely on silver solder to ensure good joints between all the "feet" and the firebox wrapper.

The girder crown stay which is not connected to the outer wrapper at all has been used in models from time to time, though only in round-top fireboxes. Based on certain designs of traction engine boilers, these stays may be made from castings or built up, and were sometimes screwed to the crown of the firebox. But this scheme cannot be recommended unless the stays are silver soldered to the crown. The author has recently built a one inch scale round-top locomotive boiler where the crown stays were made from thick copper sheet, bent into angle sections, and cut away on the lower edges as described above. The angles so formed were then silver soldered to the crown. To reduce the pressure which would otherwise have fallen upon the foundation

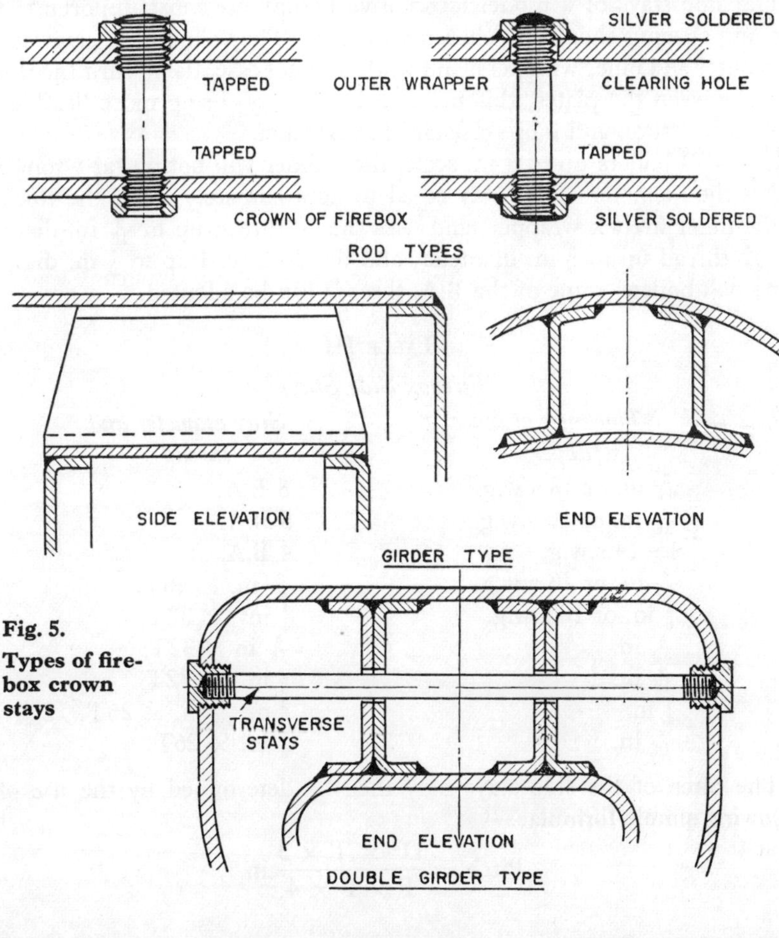

Fig. 5. Types of firebox crown stays

ring, the highest longitudinal row of side stays were made much stronger than usual—$\frac{1}{4}$ in. diameter drawn bronze rod, and these were silver soldered to both inner and outer wrappers. The boiler proved a good steamer and stood a hydraulic test pressure of 240 lb. without any trouble. It also proved a very easy boiler to build, as the rather difficult operation of silver soldering the usual form of crown stays to the outer wrapper was eliminated.

Coming to the last class of crown stay, the "sling" girder type uses a cast or malleable-iron girder screwed to the crown, a tee-shaped bar riveted to the outer wrapper, and links connecting the two by means of short pins (Fig. 6). Such stays are more difficult to make. They are used in round-top fireboxes but are really only suitable for large models where there is sufficient space between the inner and outer wrappers for easy assembly of the parts.

Although rather cumbersome in fitting up, the sling stays have one advantage in that they are much more flexible than the other types described.

Side Stays

The side stays of a model locomotive firebox are most important; their size and spacing should be chosen with care. They should be screwed into both outer and inner wrappers, and while it is not essential to turn the threads away between the plates, this makes a better stay, being more flexible, less likely to fracture, and more resistant to corrosion.

In model boilers up to 3 in. scale, the builder will not go far wrong if he makes the diameter of the stay equal to approximately twice the thickness of the inner firebox wrapper, and uses the 40 thread up to $\frac{3}{16}$ in. diameter, the 32 thread up to $\frac{3}{8}$ in. diameter, and the 26 thread up to $\frac{1}{2}$ in. diameter. For small boilers, some of the B.A. threads can be adopted.

TABLE III

Firebox Side Stays

Thickness of inner wrapper	*Stay diameter and thread*
$\frac{3}{64}$ in. or 18 s.w.g.	5 B.A.
$\frac{1}{16}$ in. or 16 s.w.g.	5 B.A.
15–14 s.w.g.	4 B.A.
$\frac{3}{32}$ in. or 13 s.w.g.	$\frac{3}{16}$ in. × 40T.
$\frac{1}{8}$ in. or 10 s.w.g.	$\frac{1}{4}$ in. × 32T.
$\frac{5}{32}$ in.	$\frac{5}{16}$ in × 32T.
$\frac{3}{16}$ in.	$\frac{3}{8}$ in. × 32T.
$\frac{1}{4}$ in.	$\frac{3}{8}$ or $\frac{7}{16}$ in. × 26T.
$\frac{5}{16}$ in.	$\frac{1}{2}$ in. × 26T.

The pitch of the side stays may then be determined by the use of the following simple formula:—

$$\text{Pitch} = \frac{D^2 \times T \times 3}{P \times F \times 4} \text{ in.}$$

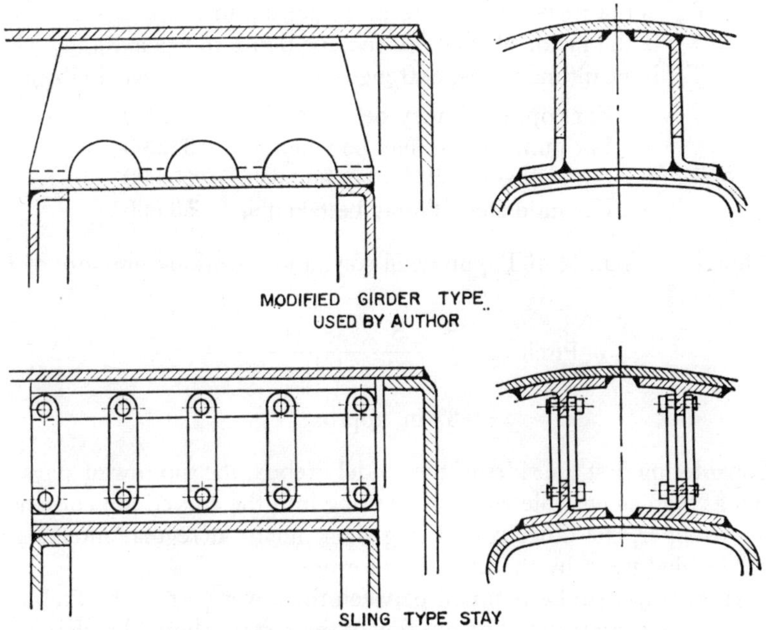

Fig. 6. Further types of firebox crown stays

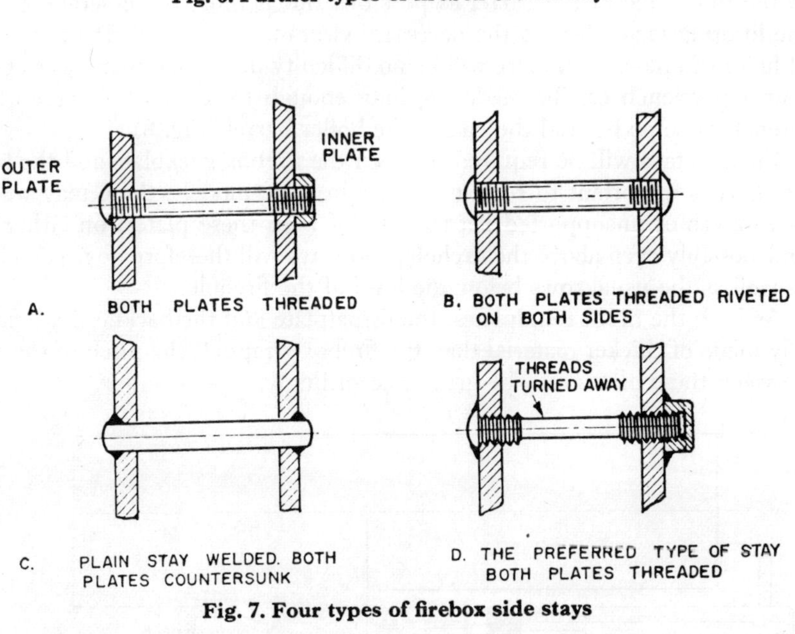

Fig. 7. Four types of firebox side stays

where D is the minimum diameter of the thread in inches (where the stay is threaded throughout, the root diameter is taken).

P is the working pressure in lb. per sq. in.

F is the factor of safety (a reliable figure to use is 8).

T is the ultimate tensile strength of the material used. in lb, per sq. in.

For copper, T may be taken as	25,000
For gunmetal T may be taken as	38,000
For monel metal T may be taken as	60,000
For mild steel T may be taken as	60,000

Thus for a $\frac{3}{16}$in. × 40T. gunmetal stay, and a working pressure of 100 lb. per sq. in.—

$$\text{Pitch} = \frac{5 \times 5 \times 38{,}000 \times 3}{32 \times 32 \times 100 \times 8 \times 4} \text{ in.}$$

$$= 0{\cdot}87 \text{ in. approx.}$$

Considering first the sides of the model firebox, the top row of stays should be put as high as possible without breaking into the curved top corners of the inner wrapper, the other rows being then put in at regular intervals at the pitch decided upon by the preceding formula.

Further stays will be required between the lower part of the firebox tubeplate and the throatplate. The top row of these stays should be placed as near to the underside of the barrel as possible, but it must be remembered that the limiting factor here is the necessary clearance for a drill. If the top row of holes can be drilled, there will be no difficulty in tapping them, as a special long tap wrench can be made up, long enough to allow the handle of the wrench to come beyond the end of the boiler barrel (Fig. 8).

Finally stays will be required between the firebox backplate and the backhead. It is important to remember that in large models, there may well be large areas of unsupported flat surface (of both these plates) on either side and possibly even above the firehole, and stays will therefore be needed here, as well as the usual rows below the level of the firehole.

As both the firebox endplates, the throatplate and the backhead are normally made of thicker material than the firebox wrapper, the pitch of the stays between these plates may be greater accordingly.

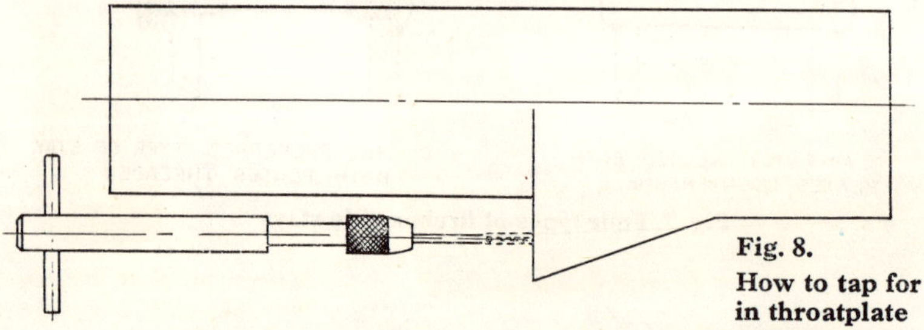

Fig. 8.
How to tap for stays in throatplate

Plate 12

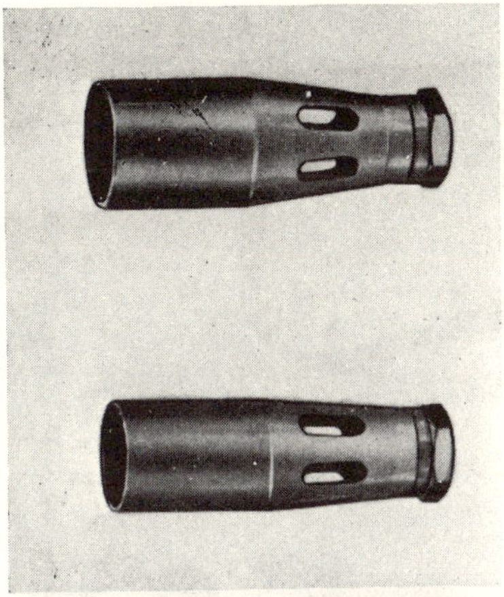

Two of the larger Sievert propane burners.

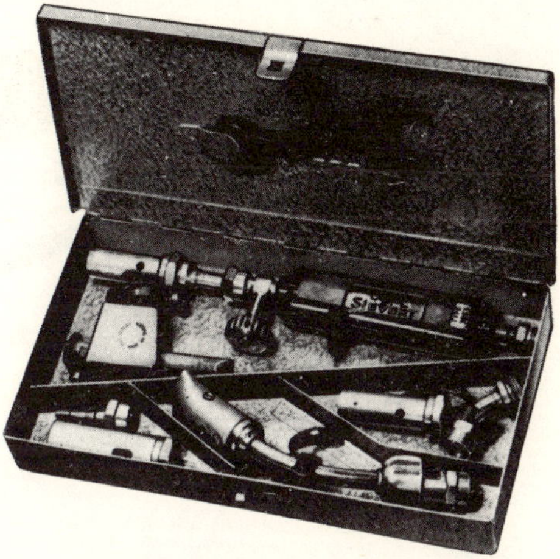

Sievert "Handyman" set for use with propane.

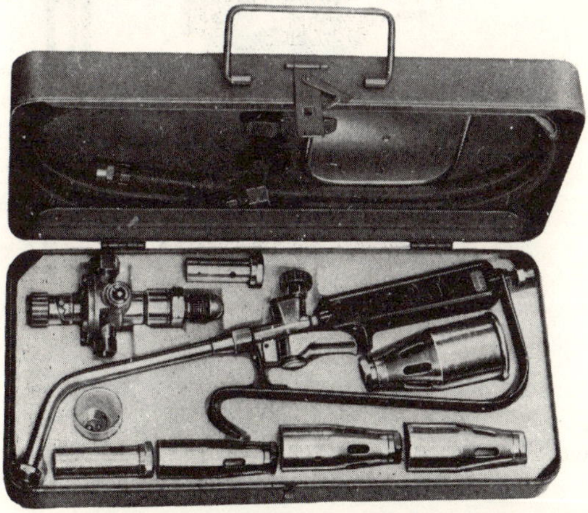

Sievert "Craftsman's service" set.

Sievert propane "hobby set" No. 2920.

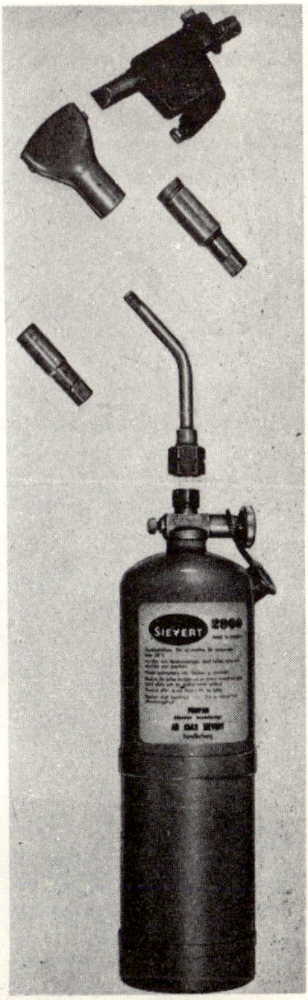

Sievert propane container No. 2960 with accessories.

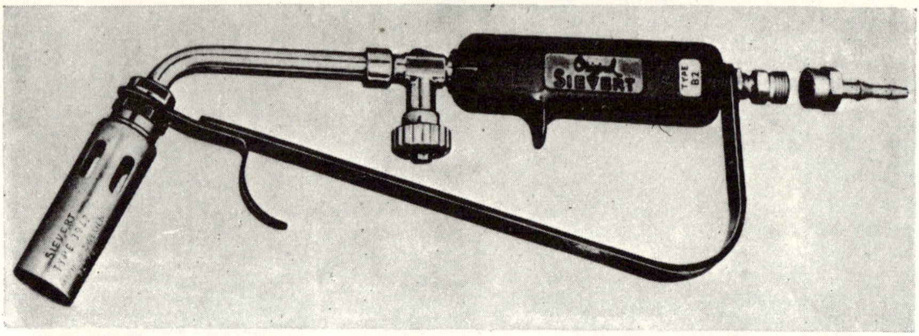

Sievert propane burner and support B.2. A useful type for the boiler maker

Right: Two petrol blowlamps by Sievert of Sweden

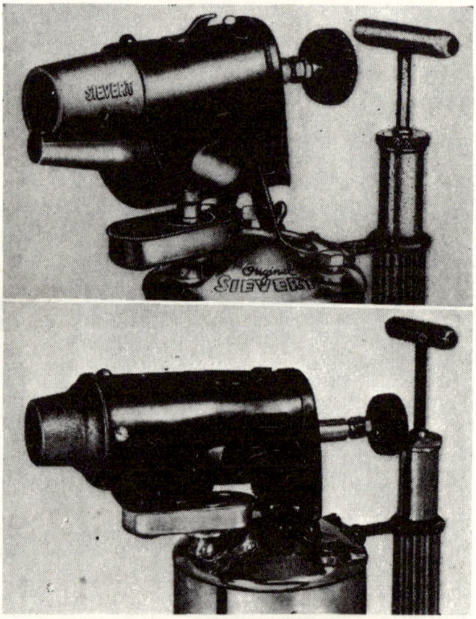

Below: A propane gas cylinder by Sievert, suitable for small boiler work

Below: A paraffin blowlamp by Sievert: photographs by Wm. A. Meyer Ltd

Three boilers by A. F. Farmer—A 5 in. gauge "Springbok" boiler, a boiler for "Boxhill" and a small gauge I boiler

Right: The author's 5 in. gauge "Firefly" boiler under construction

Below: Parts of a Belpaire firebox for a 3½ in. gauge locomotive by H. C. Henley: Photograph by G. M. Cashmore

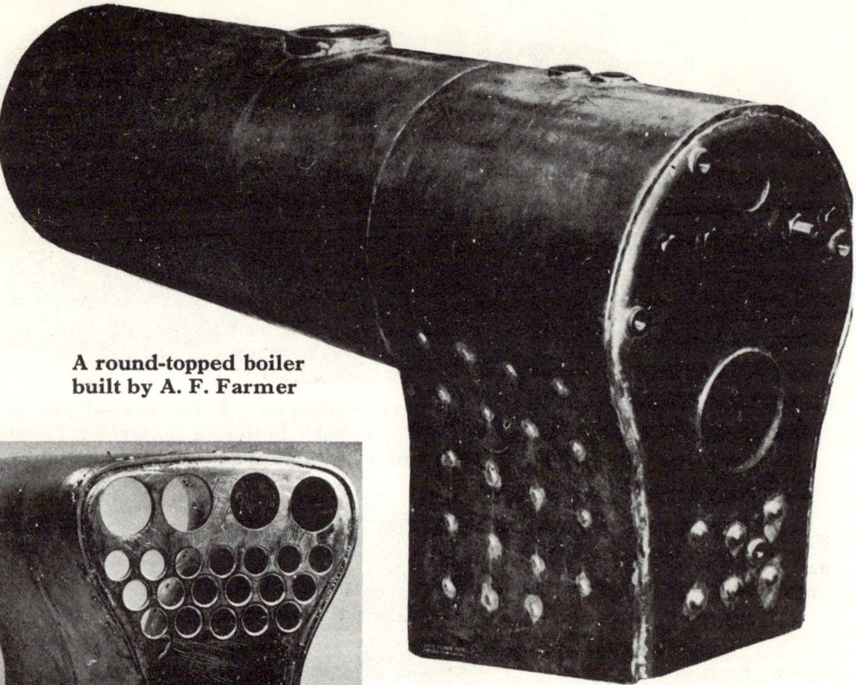

A round-topped boiler
built by A. F. Farmer

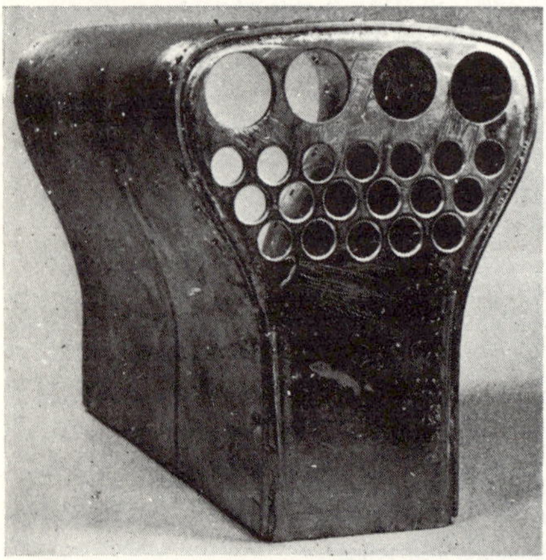

Left: The firebox of the author's 5 in. gauge "Firefly" boiler

Boiler for $3\frac{1}{2}$ in. gauge Caledonian "Dunalastair" class locomotive, built and photographed by G. M. Cashmore

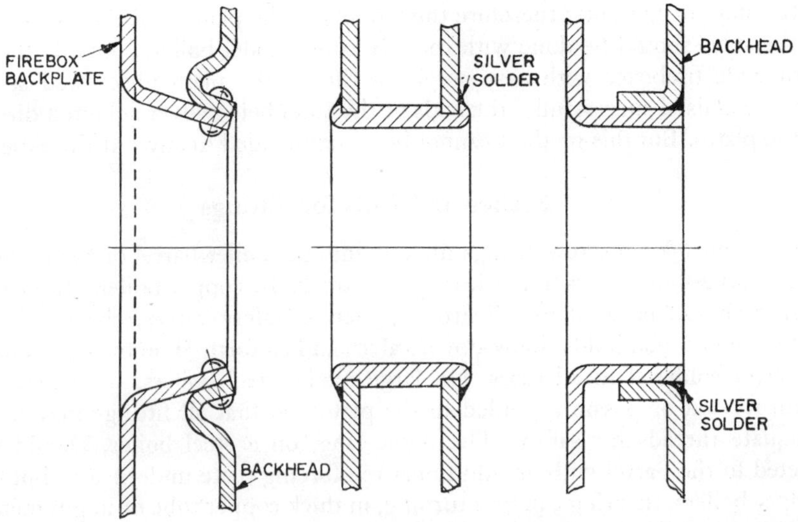

Fig. 9. Three types of firehole; that on the right designed by Mr. K. N. Harris

All side stays are best fitted with nuts on the inside of the firebox. If the stays are of copper, the heads on the outside may be riveted over. (If copper stays are turned from round copper rod, the end which is to be riveted should be annealed before fitting.) Copper, gunmetal and monel metal stays can be run over with silver solder or a high-melting point soft solder. Steel stays are not normally riveted, and if the threads are close-fitting, they should not need running over with silver solder or other alloys.

For the best results, the nuts on the inside of the firebox should be fairly thick, and in fact of such a thickness that, when screwed home, only about two-thirds of that thickness is engaged with the stay.

Belpaire fireboxes require transverse stays to support the flat sides of the outer wrapper above the level of the inner firebox. These stays may be made to a diameter equal to $1\frac{1}{2}$ or twice the thickness of the outer wrapper. They are generally fitted by means of threaded nipples as described for longitudinal stays, though in copper boilers they can be screwed direct into the wrapper on one side of the boiler, nutted over on the other side, and then silver soldered.

In boilers up to one inch scale, one row of transverse stays is generally sufficient: in large models two rows will probably be required, according to the area involved.

Builders of large model Belpaire fireboxes should not forget the top front corners of this type of firebox, together with the opposite corners of the backhead. These corners may well project so far out from the circular barrel that they leave a large flat area unsupported by stays. A single longitudinal stay from the front corners to the backhead should overcome this difficulty.

The important point to remember about all boiler stays is that the threads in the plates are intended to take the various stresses involved, not the heads of the stays or the nuts; therefore the tapping of the plates and the threading of the stays should be done with care. In some model boilers, notably those with wide fireboxes with easy access to the inside, plain unscrewed stays have been used successfully, the ends of the stays being silver soldered direct to the plates. But this method cannot be recommended to any but the expert.

Bushes and Pads for Fittings

It is a mistake to screw fittings directly into the boiler barrel or the firebox plates, unless the plates are at least $\frac{3}{16}$ in. thick. In copper boilers, threaded bushes should be used for all fittings, such as safety valves, check valves, water gauges, manifolds, blow-down valves and so forth. The bushes should be silver soldered in all cases. On large model steel boilers, pads may be used, riveted or possibly welded to the plates, so that all fittings may have adequate threads for safety. The dome ring, on a steel boiler, should be riveted to the barrel with an additional reinforcing plate underneath, but on copper boilers, this ring can be a turning, in thick copper tube or in gunmetal, and silver soldered to the barrel.

The Firehole Ring

For copper locomotive boilers, a turned firehole ring is recommended. This saves having to flange either the firebox backplate or the backhead. The best material to use is thick-walled copper tube, which is obtainable commercially up to a size large enough for the biggest model locomotive boiler.

The ring is turned down at each end to form a step in the middle. One end is then inserted in a hole cut in the firebox backplate. It is then flanged over and brazed. When the backhead is finally assembled, the ring is flanged over on the outside of the boiler, thus making a very strong job. If an oval firehole is preferred, the turned ring can be squeezed oval in the vice before assembly, and the firebox plates shaped to suit.

In large model steel boilers, a turned ring is subject to some disadvantages:— a relatively large mass of metal is concentrated to an undesirable extent, and as the area of the ring actually in contact with the water is small, the heat is not transferred sufficiently rapidly, thus the plates in the immediate vicinity tend to burn and develop cracks. The construction is somewhat rigid, thus movement due to expansion is restricted. The length of the rivets used may make maintenance of steam tightness difficult.

To avoid a turned ring in steel boilers, the plates of both the inner and outer fireboxes may be flanged outwards to form a joint. In this scheme (Fig. 9) the firebox backplate should be flanged as little as possible and to an easy radius. The only drawback to this design is the fact that scale tends to lodge in the corners, between the two plates.

Another method that can be used in large model steel boilers is to dish the firebox backplate towards the backhead, and to use a ring of corresponding reduced thickness.

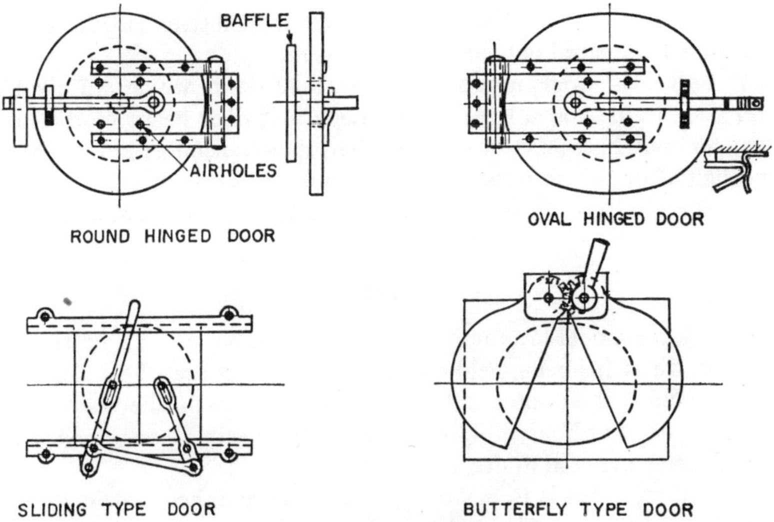

Fig. 10. Firehole doors

Firehole Doors

Three classes of firehole doors are used on model locomotive boilers. The plain hinged door, with a vertical hinge on the left or right-hand side of the door; the sliding type door, and the "butterfly" door.

A door which is hinged at the top is most undesirable on models below 3 in. scale, as it prevents the driver from seeing the fire properly. A door hinged at the bottom is not usually satisfactory owing to the tendency for the hinge to become jammed with small particles of coal or ash.

The side hinged door is easy to make and satisfactory in service. The handle should be made long enough to enable the driver to reach it easily while the engine is in motion. The latch should be designed so that the door can be opened and closed positively. On locomotives to the smaller scales, it is not a bad plan to design the firedoor so that it can be pushed shut with one end of the shovel and opened with the other.

To ensure complete combustion of the fuel, a few small air holes are usually provided in the firedoor. The "top air" allowed over the fire in this way helps to reduce smoke. A baffle can be arranged just inside the door to deflect the top air and to prevent it from being drawn straight across to the tubeplate. This however is not of great importance in small copper boilers.

The sliding type firedoor is quite satisfactory in model form and looks better than the side-hinged pattern. In model locomotives, this type of door can generally be opened and closed by a judicious tap with the firing shovel, a great convenience when firing on a run. One disadvantage of the sliding firedoor is the tendency for the lower runner to become choked by coal dust. This can be overcome by leaving one side of the lower runner channel open, immediately below the firehole. Alternatively, the lower edges of the two doors can be formed into a channel section and a plain strip used for the lower runner, suitably spaced out from the backhead.

The "butterfly" firedoor consists of two half doors which are pivoted to a main door frame above the firehole. Each half door has a toothed sector at its upper end. These sectors gear together, thus transmitting any movement of one half door to the other. Although more work than other designs of firedoor, the "butterfly" type is very suitable for model locomotives.

Firegrates

Firegrates for model locomotive boilers are sometimes built up from mild steel strips, the strips being held together by two rods near the ends, the rods being threaded at their outer ends, and turned spacers put between each bar.

Grates made of ordinary mild steel do not, however, last very long; rustless steel is a better material to use.

An even better plan is to make a pattern and have the whole grate cast in iron, as once the pattern is made, further castings can be quickly obtained as the grates become burnt out. The firebars of a cast grate can also be made to the correct shape, tapered on both sides towards the bottom, so that ashes do not tend to stick between the bars, but fall through into the ashpan.

The fitting of the grate to the firebox should be carried out with care, as it is useful to be able to drop it quickly in emergencies. The grate should not in any case be attached permanently to the firebox; it can generally be located by filing notches in the foundation ring, into which are fitted extension lugs cast on the grate. Where wide fireboxes are concerned, it will often be found that it is not possible to get the grate into position if it is made in one piece. It may therefore be made in two halves, with suitable lugs and recesses cast on the two fitting edges, so that location is simplified. Another method is to make the grate in three parts, pivoted at the front end, the middle section being allowed to swing, when required, so as to deposit the ashes into the ashpan.

For locomotives of $\frac{1}{2}$ in. scale, the firebars may be made approximately $\frac{3}{32}$ in. wide at the top and about $\frac{5}{16}$ in. deep. For $\frac{3}{4}$ in. scale grates, a firebar section of about $\frac{1}{8}$ in. by $\frac{3}{8}$ in. is generally satisfactory, with $\frac{5}{32}$ in. air spaces. $\frac{3}{16}$ in. by $\frac{1}{2}$ in., with $\frac{7}{32}$ in. air spaces will be found successful for 1 in. scale boilers, and for $1\frac{1}{2}$ in. scale locomotives, the bars may be made $\frac{1}{4}$ in. by $\frac{5}{8}$ in. or $\frac{3}{4}$ in. with $\frac{9}{32}$ in. air spaces.

Ashpans

All locomotive boilers burning solid fuel must be fitted with ashpans. A properly designed ashpan helps the fire to burn evenly over its whole area, as well as preventing hot ashes from falling on to the track and possibly damaging the sleepers.

The shape of the ashpan depends largely on the proximity or otherwise of driving or coupled axles, or in the case of locomotives with wide fireboxes such as most Pacifics and Atlantics, the proximity and arrangement of the trailing axle. Generally speaking, the best results are obtained if the front of the ashpan is completely closed, and the rear end left open for the ingress of air. If the grate has a bend in it, the bottom of the ashpan should where possible follow the contour of the grate.

If the ashpan has to be shallow, due to the close proximity of coupled axles, additional air space should be provided.

On large model locomotives, dampers may be fitted, hinged on the horizontal centre line, so as to secure a more even distribution of the air admitted. The dampers are controlled by rods leading up to levers inside the cab.

Ashpans are generally built up from mild steel sheet, but rustless steel is a much better material, owing to the rust problem: ashpans tend to get wet from the boiler water. A suitable thickness for the ashpan would be 16 or 18 s.w.g. for a $\frac{3}{4}$ in. scale engine or 13 s.w.g. ($\frac{3}{32}$ in.) for up to $1\frac{1}{2}$ in. scale. The joints are welded, brazed, or angled and riveted. A bottom door, to allow the ashes to drop through between the rails without removing the ashpan itself, is a useful fitting. The pivot of such a door must be protected from the ashes by a ledge arranged just above it, otherwise the door will tend to jam in service.

Both grates and ashpans should be made an easy fit and it should be possible on practical working engines to drop both of them quickly in emergencies.

Combustion Chambers

Where the model locomotive boiler has, of necessity, a long barrel, and particularly if it has a wide firebox, a combustion chamber generally improves steaming through its big increase in what is virtually firebox volume. As the combustion chamber must be made to the same shape as that part of the firebox tubeplate to which the tubes are normally attached, some form of staying is necessary, unless the material used in its construction is very thick.

Fig. 11. A useful type of ashpan

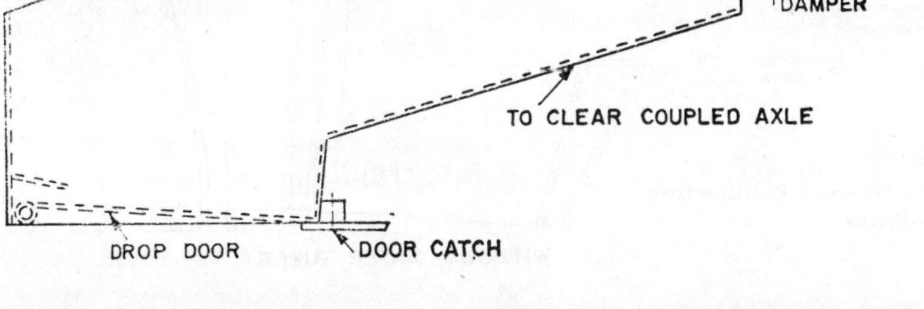

As a rough guide, it may be said that a combustion chamber in a $\frac{3}{4}$ in. scale boiler which, if properly stayed, would be made from 15 to 16 s.w.g. material, would have to be made from 8 s.w.g. metal if unstayed.

The usual solution to the staying problem is to fit water tubes, set almost vertically at required intervals, and generally in two rows. These water tubes should not be less than $\frac{1}{2}$ in. diameter in a $\frac{3}{4}$ in. scale boiler, or $\frac{5}{8}$ in to $\frac{3}{4}$ in. diameter in a 1 in. scale boiler, and should be made of fairly thick material.

But many model locomotive designers no longer favour the use of combustion chambers in the smaller scales, owing to their inaccessibility. It will be appreciated that once the inner firebox assembly, complete with the combustion chamber and tubes, is silver-soldered into position inside the boiler, it is impossible to reach the tubeplate should a leak occur through faulty workmanship or bad soldering or should a leak develop when the engine is in service.

If the builder decides to use a combustion chamber, the author recommends that a thorough hydraulic test be carried out on the centre firebox assembly before it is erected in the boiler. This, of course, is not by any means a straightforward task, but it can be done by blanking off the outer

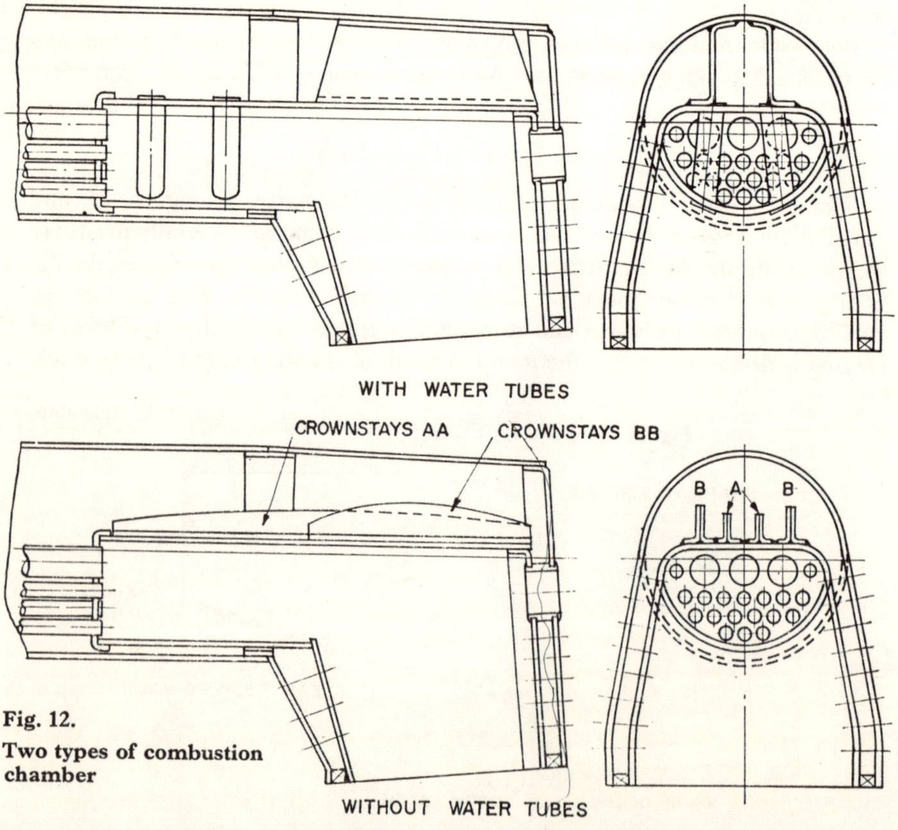

Fig. 12.
Two types of combustion chamber

ends of every tube, the firehole and the open bottom of the firebox. A stout plate, with a washer of thin rubber or similar material, could be used across the firehole, with a long rod threaded at each end to run through a convenient tube or flue to the front end. A similar arrangement could be clamped to the underside of the firebox by long bolts and crossbars top and bottom.

On the other hand, if the builder prefers to do without a combustion chamber in a long barrelled boiler, efficiency may be maintained by the use of larger diameter tubes and flues, by setting the smokebox tubeplate further into the barrel (thus decreasing the effective length between tubeplates) and by sloping the firebox tubeplate forwards, so as to bring the upper part of the tubeplate to which the tubes are attached, closer to the front of the boiler.

Thermic Syphons

There is no doubt that the advantages shown by the fitting of thermic syphons to full-size locomotive boilers can be realised in the smaller scales. The feature that the crown of the firebox is always covered by water so long as the level does not fall below the level of the bottom of the syphon, is perhaps the most valuable one. But with the greatly improved water circulation brought about by the syphon, trouble may be encountered with priming or with faulty water gauge readings. Certainly it is worthwhile siting the safety valves and the regulator port well away from the top of the syphon and it may prove necessary to fit the water gauge to a column rather than direct to the boiler.

Brick Arches

Although brick arches are desirable fitments in larger scale model locomotives, it is doubtful if they have much effect in boilers below 1½ in. scale. They are also rather difficult to locate in small boilers and may impede the operation of tupe sweeping. If used, they may be made from some form of fire-brick and could be supported at the desired angle by extensions screwed on to some of the firebox side stays in place of the usual nuts.

Brick arches made of some suitable refractory material are recommended in steal boilers of 1½ in. scale and upwards.

Fusible Plugs

The question of whether to fit a fusible plug to a model locomotive boiler is a debatable one. If the boiler has been made by riveting and soft soldering, then a fusible plug of some kind is essential. If the boiler is one which has been brazed or silver-soldered, but contains soft solder in the form of caulking on the stays, then a fusible plug is certainly desirable.

If, however, the boiler is one which has been entirely brazed or silver-soldered, a fusible plug is not essential and could be omitted. Two types of such plugs are shown in Fig. 13 and it is probable that the type containing

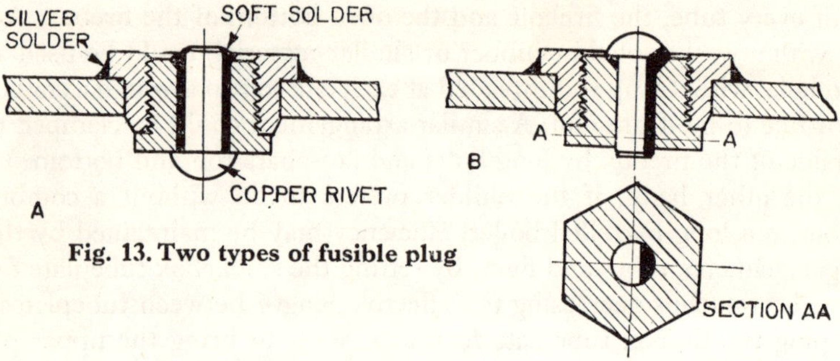

Fig. 13. Two types of fusible plug

a loose copper rivet, held only by the solder, is the better of the two. The fusible plug should never be screwed direct into the plate of the firebox crown in small boilers, but a properly turned bush should be silver-soldered into the crown and arranged in a convenient place, so that the fusible plug can be screwed in through the firebox.

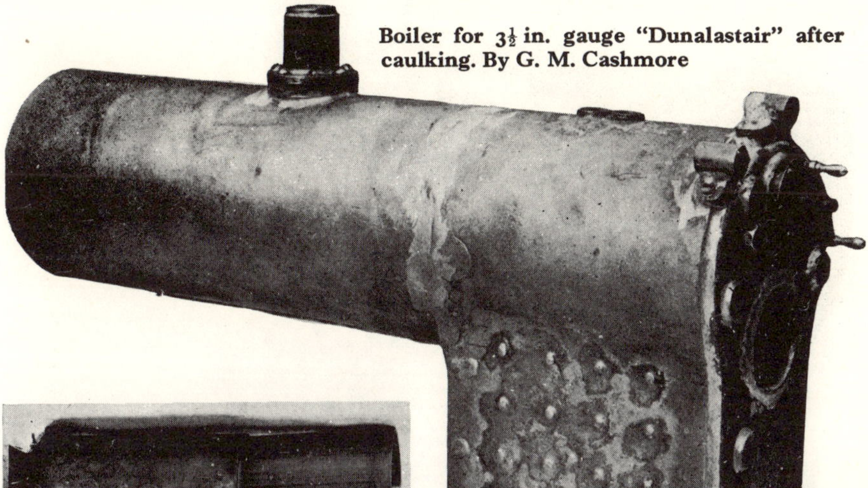

Boiler for 3½ in. gauge "Dunalastair" after caulking. By G. M. Cashmore

Building up a two-piece boiler barrel

Right: Inner firebox and tube assembly

Below: A superheater complete with "wet" header

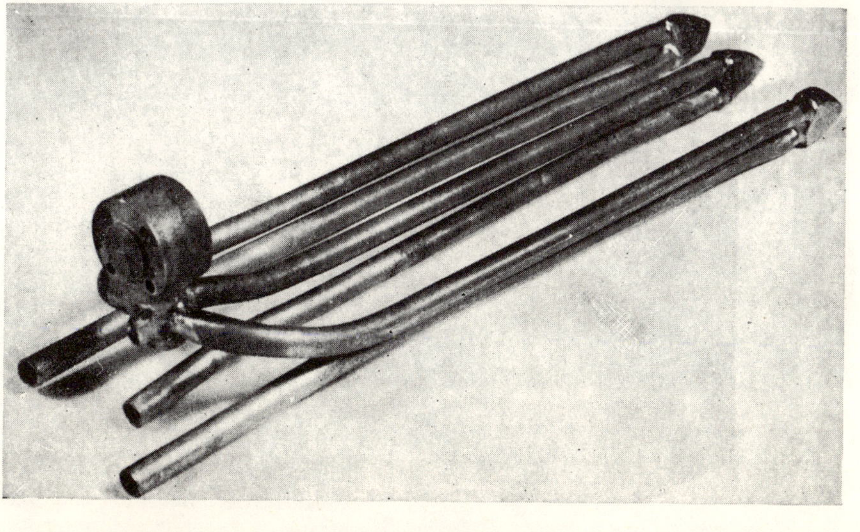

Plate 17

Backhead view of the author's 5 in. gauge "Firefly" boiler

Parts of the author's "Simplex" boiler built by Clive Schroder

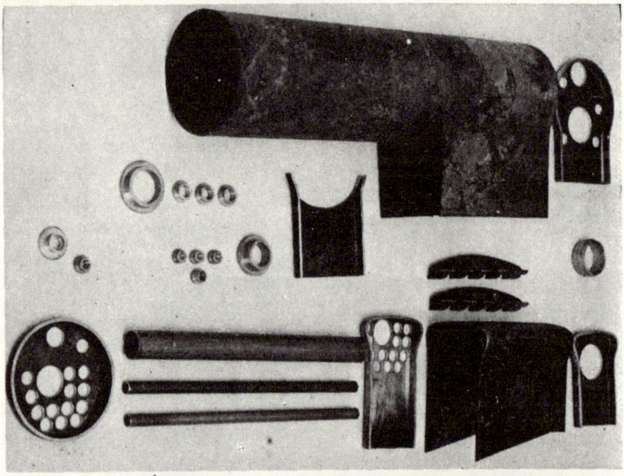

Below left: "Simplex" inner firebox and tube assembly

Right: Firebox for 5 in. gauge "Terrier"

5 in. gauge "Terrier" boiler ready for fitting of backhead

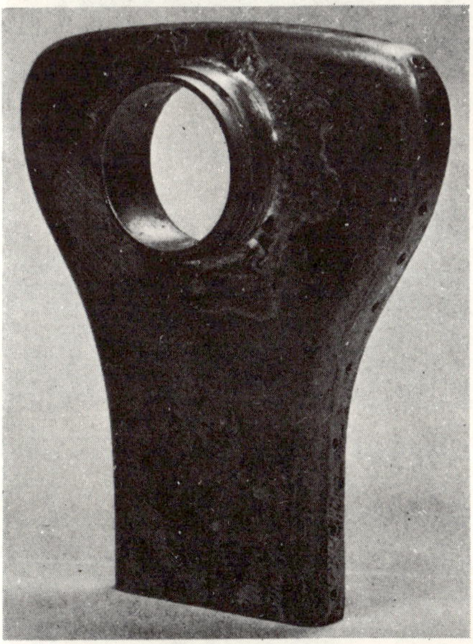

Firebox wrapper ready for assembly

Left: Firebox backplate

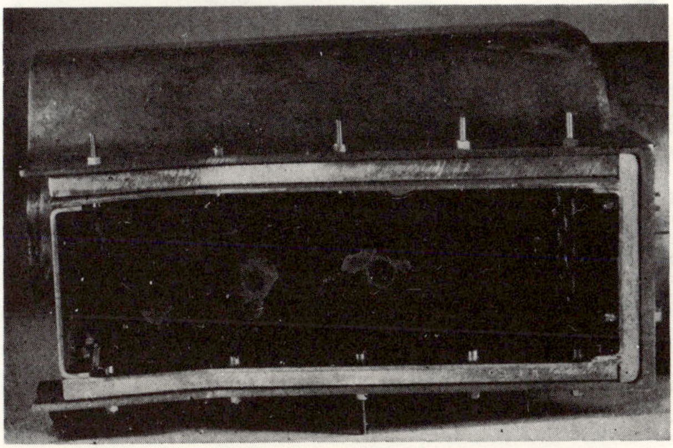

Fitting the foundation ring to a $3\frac{1}{2}$ in. gauge boiler

Cab view of Mr. R. W. Fenwick's 5 in. gauge "Terrier". Note neat backhead fittings

Bending rolls made by the author

CHAPTER THREE

Welding and Cutting

A WIDE range of processes are today available for welding steel, stainless steel and copper, the three metals most used in boiler work. The processes can be divided into two main sections: Fusion welding, or welding without pressure, which includes Gas welding processes, Arc welding, Thermit welding and Bronze welding. Welding with pressure, which includes Resistance welding processes, Forge welding, Stud welding and Thermit welding with pressure.

In most forms of arc welding, an arc, which is a low voltage high current discharge, is formed between the work to be welded, which is connected to one pole of the electrical system, and an electrode held in a suitable holder which is connected to the other pole. The heat of the arc raises the workpiece at the point of welding and the electrode to a very high temperature (over 3,000 deg. C.), the higher temperature—in the case of direct current—being at the positive pole. The electrical system may be D.C. or A.C., depending on the process, material and welding technique to be used, and is supplied by a generator or transformer specially designed for welding.

Although some types of arc welding are suitable for model boiler work, the gas welding process is more useful for the amateur, especially when the equipment required can be used for other purposes such as the cutting of sheet metal.

Oxy-Acetylene Welding and Cutting

Of all the Oxy-fuel gas welding processes, oxy-acetylene welding is by far the most widely used. It is a very versatile welding process and is ideal for much of the work involved in the construction of model boilers. Acetylene is not always used as the fuel gas for welding and cutting purposes, the choice of the fuel gas to be used depending on the material being welded, the welding temperature required and its cost. Table IV gives an indication of the flame temperatures of the various flames, and the appropriate calorific values of the fuel gases.

Acetylene is normally supplied under a pressure of 225 lb. per sq. in., in cylinders filled with a porous material containing acetone which holds the gas in solution. The cylinders are made of solid-drawn steel and can be obtained in various sizes. They should always be used in the upright position, otherwise the contents of the cylinder which are liquid are likely to reach the outlet valve.

In order to distinguish acetylene cylinders from other cylinders containing gas, thay have maroon-coloured markings, and the outlet valves have left-hand threads. Calculation of gas consumption is usually based upon measurements of the weight of the cylinders before and after use, rather than upon pressure gauge readings. One pound of acetylene gas is approximately equivalent to 14·5 cu. ft.

Oxygen is generally obtained in cylinders, compressed to a pressure of about 1,980 lb. per sq. in. The cylinders are identified by black paint markings, and the outlet valves have right-hand threads.

TABLE IV FLAME TEMPERATURES

Type of flame	*Approx. flame temperature (Deg. C.)*	*Approx. Calorific value of fuel gas B.Th.U. per cu. ft.*
Oxy-acetylene	3,200	1,450
Oxy-propane	2,500	2,500
Oxy-hydrogen	2,200	345
Oxy-coal gas	2,000	425–625

Welding and Cutting Equipment

The equipment required for oxy-acetylene welding normally includes the following items:

1. Oxygen cylinders.
2. Acetylene cylinders.
3. Regulators and gauges.
4. Welding blowpipe and nozzles.
5. Cutting attachment and nozzles.
6. Hoses.
7. Welding rods and fluxes.
8. Ancillary equipment such as goggles, wire brushes, spanners, trolleys and stands, gloves and aprons.

The amateur or professional model boiler maker who wishes to be properly equipped to deal with all types and sizes of boilers cannot do better than to acquire a complete kit such as the "Saffire 3", manufactured by the British Oxygen Company. The "Saffire 3" combined outfit includes the following items: A two-stage oxygen regulator complete with pressure gauges, fifteen ft. of $\frac{5}{16}$ in. hose for oxygen, fifteen feet of $\frac{5}{16}$ in. hose for acetylene, a two-stage acetylene regulator complete with pressure gauges, seven sizes of welding nozzles, welding blowpipe, welding goggles with spare lenses, cutting nozzles in three sizes, cutting attachment for blowpipe, spindle key, head nut spanner, outfit spanner, regulator spanner, data card, spark gas lighter and metal carrying case.

The "Saffire 3" blowpipe is capable of welding mild steel from 20 s.w.g. up to $\frac{5}{16}$ in. thick, or copper and stainless steel up to $\frac{1}{2}$ in. thick. The cutting attachment will cut mild steel sheet from 20 s.w.g. up to 2 in. thick. It incorporates the principle of nozzle mixing; this gives the blowpipe better handling qualities and greater safety. The attachment has its own heating oxygen valve and cutting oxygen lever. The head is made of monel metal, the gas tubes are in stainless steel and the handle extruded aluminium. Either acetylene or propane can be used in the cutting attachment with the appropriate nozzle. A selection of spade and roller guides and circle cutting attachments can also be supplied by the manufacturers.

For those who only intend to build small boilers such as locomotive boilers up to half-inch scale or other boilers of comparable size the "Saffire lightweight" welding blowpipe or the "Firefly" ultra-lightweight combined set may be considered. the "Saffire lightweight" is designed for welding mild steel sheet from 20 s.w.g. up to $\frac{5}{32}$ in. thick, or copper up to $\frac{1}{8}$ in. thick. The total weight of this blowpipe is only ten ounces so that fatigue in use is reduced to a minimum. The "Firefly" combined set is capable of welding mild steel from 20 s.w.g. up to $\frac{1}{4}$ in. thick, or copper up to $\frac{1}{8}$ in. thick; it is also able to cut mild steel up to $\frac{3}{4}$ in. thick.

Both the "Saffire lightweight" and the "Firefly" blowpipes use hose of $\frac{3}{16}$ in. internal diameter. Only the correct type of hose should be used, fitted with hose protectors. The B.O.C. "Alda" gas hose contains 40 per cent rubber, and is manufactured by wrapping the rubber-containing lining over a mandrel, and winding on to it cotton tape or duck; it is then given a protective outer coating, vulcanised, and the mandrel extracted. This method of manufacture keeps the hose straight in service and prevents kinking. The oxygen hose is always coloured black and the acetylene hose red. Before any hose is connected up, it should be blown through to remove dust; new hose may contain chalk dust, which should be carefully blown out.

Operation of Oxy-acetylene Equipment

1. Before attempting to light a blowpipe, the job should be prepared and set up in a well-ventilated place, and fluxed if appropriate, the welding or brazing rods or strips laid ready to hand. The operator should be suitably dressed and be wearing the proper goggles.

2. Both the oxygen and the acetylene cylinders must be stood up vertically, this can best be done by using the proper manufacturers' stands.

3. The jointing surfaces in cylinder valves and regulators are next examined, to ensure that they are clean and free from oil and grease.

4. The valve on the oxygen cylinder (the cylinder painted black) can now be opened momentarily, in order to dislodge dirt or obstructions in the valve, and then closed.

5. The valve on the acetylene cylinder (the cylinder painted maroon) is also opened momentarily and then closed.

6. The oxygen regulator (painted black) is screwed into the oxygen cylinder valve. Right-hand screw threads are used on the oxygen cylinder valve outlet and regulator connection.

7. The acetylene regulator (painted maroon) is screwed into the acetylene cylinder valve. Left-hand screw threads are used on the acetylene cylinder valve outlet and regulator connection. The regulators should be tightened sufficiently to ensure that the joints are gas-tight, but excessive force must not be used.

8. Blow through the hoses to remove dust and connect them to the regulator outlets by means of the screwed connections. The oxygen hose is coloured black and the acetylene hose maroon.

9. The other ends of the hoses, those fitted with the hose protectors, are now connected to the blowpipe, the acetylene hose to the connection marked A, and the oxygen hose to the connection marked O. The blowpipe control valves should be kept closed. A high or low pressure blowpipe can be used with Dissolved acetylene. If a low-pressure blowpipe is used on dissolved acetylene, the acetylene pressure should never exceed 2 lb. per sq. in.

10. Fit the appropriate size nozzle to the blowpipe (see Table V).

11. Open the cylinder valves, one turn only, VERY SLOWLY, by means of the cylinder key. If the valves are opened suddenly, there is danger of an accident.

12. Set the regulators at the correct working pressures (Table V).

13. Open the acetylene control valve on the blowpipe, *wait/a few seconds until any air has been blown out and pure acetylene is coming from the nozzle*, then light, preferably by means of the spark lighter.

14. Reduce or increase the acetylene supply by means of the blowpipe valve until the flame just ceases to smoke.

15. Open the oxygen control valve on the blowpipe until the white inner cone in the flame is sharply defined, with the merest trace of an acetylene haze.

The blowpipe is now ready for the welding of steel and work can be started.

On completion of the job, the following procedure should be adopted:

1. Turn off the acetylene first by means of the control valve on the blowpipe, then turn off the oxygen.

2. Close both the cylinder valves.

3. Open the blow pipe valves one at a time to release the gases left in the hoses: open the oxygen valve first, and shut it, then open the acetylene valve, and shut it.

4. Unscrew the pressure regulating screws on the oxygen and acetylene regulators.

The size of the nozzle given for a particular thickness of steel is for general guidance only, and will of course vary according to the mass of metal. For work on copper, a nozzle size two sizes larger than that shown for steel should be chosen.

TABLE V HIGH-PRESSURE BLOWPIPES

Nozzle sizes, working pressures and gas consumptions

Mild steel plate thickness (in.)	*Nozzle size*	*Regulator Pressure lb. per sq. in. Oxygen and acetylene Saffire*	*DH.*	*Approx. consumption of each gas Cu. ft. per hour*
1/32	1	2	2	1
3/64	2	2	2	2
1/16	3	2	2	3
3/32	5	2	2	5
1/8	7	2	2	7
5/32	10	3	3	10
3/16	13	3	4	13
1/4	18	3	7	18
5/16	25	4	9	25
3/8	35	4	–	35
1/2	45	5	–	45
3/4	55	5	–	55
1	70	6	–	70

Flame Adjustment

The flame obtained as described in 15 above is known as a neutral flame. In this condition the blowpipe is burning equal quantities of oxygen and acetylene. It is generally advisable to have a slight haze of acetylene around the centre cone because in practice there is a tendency for the flame to become slightly oxidising as welding proceeds.

If an oxidising flame is required, the acetylene control valve on the blowpipe is closed until the required flame is obtained.

If a carburising flame is required (i.e. a flame using excess acetylene), the acetylene control valve is opened, and a feather of acetylene produced at the end of the central white cone. A carburising flame is only used for hard facing and building up worn surfaces on steel or cast iron, so should not be required by the boiler maker.

Welding Rods and Fluxes

The correct type of welding rod must be chosen according to the metal to be welded (see Table VI). No flux is required for the welding of mild steel, but the appropriate flux must be used in the case of stainless steels, copper and copper alloys.

When heated metal is exposed to the atmosphere, it combines with the oxygen, and metallisc oxides form on the surface. In the case of welds, the oxides tend to mix with the metal, causing it to become porous and weak. The fluxes which are added during welding reduce and remove these oxides by the formation of a fusable slag of low specific gravity. The slag absorbs the oxides and floats on the surface of the molten metal. In the case of mild steel, the oxide is lighter than the molten metal so that it floats to the surface without the use of a flux.

During welding, gases are released, and these can become trapped as the metal solidifies, thus causing blowholes. If a correctly balanced flux is used, the gases are released before they can become trapped; the flux also increases the fluidity of the molten metal, so that the possibility of blowholes is reduced.

The melting point of the flux must be lower than the melting point of the metal, yet not too low or the flux may be melted and driven off before doing its principal job. The flux must also spread freely over the surface of the heated metal to ensure complete oxide absorption.

TABLE VI ALDA WELDING RODS AND FLUXES

Welding rod	*Applications*	*Sizes (ins.)*	*Ult. tensile stress tons-sq. in.*	*Brinell hardness*	*Meltg. point C.*	*Flux*
Low carbon mild steel copper-coated	High-grade welding, of mild steel or wrought iron	$\frac{3}{64}$ $\frac{1}{16}$, $\frac{5}{64}$, $\frac{3}{32}$, $\frac{7}{64}$, $\frac{1}{8}$ and larger	25·0	121	1,490	None
Low carbon mild steel	General welding of mild steel	$\frac{1}{16}$. $\frac{1}{8}$, $\frac{3}{16}$,	25·0	121	1,490	None
Stainless steel-decay resistant (Niobium bearing)	Welding austenitic stainless steel	$\frac{1}{16}$, $\frac{3}{32}$, $\frac{1}{8}$, $\frac{3}{16}$,	35·0	175	1,440	Botectic
Stainless steel-decay resistant (Molybdenum bearing)	Welding austenitic stainless steel contg. molybdenum	$\frac{1}{16}$, $\frac{1}{8}$	—	—	1,430	Botectic
Stainless steel-heat resistant	Welding stainless and heat resisting steels	$\frac{1}{16}$, $\frac{1}{8}$	37·0	180	1,460	Botectic
Copper Silver Alloy	Welding copper (fireboxes, etc.)	$\frac{1}{16}$, $\frac{3}{32}$, $\frac{1}{8}$, $\frac{3}{16}$, $\frac{1}{4}$.	13·8	52	1,068	Copper Silver
Brazotectic. (Silicon-bronze)	Bronzewelding copper, welding brass.	$\frac{1}{16}$, $\frac{3}{32}$, $\frac{1}{8}$, $\frac{3}{16}$, $\frac{1}{4}$.	29·12	117	875	Uni-bronze

Welding Technique

The author has mentioned previously the importance of preparing the work beforehand, and of having all the necessary items that are likely to be required during welding ready at hand, including a bucket of water.

The correct method of lighting up has been described. If the blowpipe flashes back on lighting up, it is because the regulators are not set to the correct pressures, or the light has been applied before the flow of fuel gas has been properly established. If the flame snaps out when the blowpipe is in use, it is because:

1. The regulator pressure and/or gas flow are incorrect—either too high or too low.
2. The nozzle has become obstructed.
3. The nozzle is held too close to the work.
4. The nozzle has become overheated.

In the case of either a flash back or the flame going out, both blowpipe valves must be shut, the regulator settings and cylinder pressures checked, and the blowpipe re-lighted as described previously. In the case of 4. only, the nozzle and blowpipe head should be plunged into cold water. Make sure that the nozzle is tight before re-lighting.

In normal use, the blowpipe nozzle is kept cool by dipping into a bucket of cold water from time to time; in this case, a trace of oxygen should be left flowing to prevent water entering the nozzle.

Welded Joints

There are four main types of joint: Butt, fillet, corner and edge. The butt joint is employed when joining two edges together, the fillet is used when welding a vertical piece to a flat piece or securing one plate to another, the corner joint connects the edges of two plates at 90 degrees or any other angle.

The edges of the metal must be suitably shaped. If the sheet metal to be joined is less than $\frac{3}{32}$ in. thick, the edges should be bent over or flanged, so as to increase the area of contact. Sheet metal of $\frac{3}{32}$ in. thickness and above may be joined by means of a plain butt weld (see Fig. 14). A single vee butt join is used for sheet from $\frac{1}{8}$ in. to $\frac{3}{16}$ in. thick, the vee being about 80 degrees,

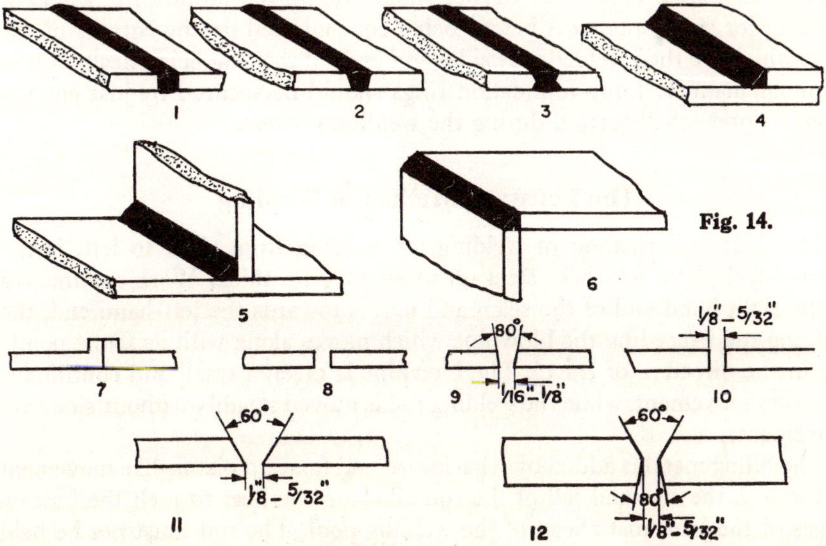

Fig. 14.

while sheet from $\frac{1}{4}$ in. to $\frac{5}{16}$ in. thick may be joined by a plain butt weld or a single vee weld, the vee having an included angle of about 60 degrees. Double vee butt joints are required where the plate thickness is $\frac{5}{8}$ in. or over.

Apart from the actual edges of the metal to be joined are quite clean, the metal should also be thoroughly cleaned of rust grease or paint, etc, for a distance of about three inches from the weld.

The Prevention of Distortion

The expansion of the heated metal during welding, followed by contraction during cooling, often causes the work to distort. Distortion is however reduced to a minimum if the parts concerned are free to expand and contract without restraint.

Distortion cannot be tolerated in model work, so steps must be taken to keep it to a minimum. In the case of plain boiler barrels, the edges of the plate can first be brought together and tack welded; i.e. single welds are first placed at regular intervals along the seam. Thin sheets should first be joined by a lap strip laid the full length of the join and riveted, using sufficient rivets to hold the job firmly. The rivets should be made of the same metal as the sheet used for the barrel. After the riveting has been completed, the joint is welded in the usual way.

Most distortion in mild steel welding is caused by the heat put into the metal adjacent to the weld. Thus a hot flame, which moves quickly, will generally cause less distortion than one of lower temperature which moves more slowly and gives the heat time to travel further from the welding point.

When welding joints on large and complex work, a certain amount of preheating is advisable. For instance when welding the foundation ring of a steel model locomotive boiler, the burner should be moved around the whole of the ring for some moments before being concentrated on the corners of the ring, which are then welded; the sides and ends of the ring being dealt with in a similar manner. Long foundation rings should be secured by just enough rivets to prevent distortion during the welding process.

The Leftward Method of Welding

The Leftward method of welding, or welding from right to left, is the normal technique for sheet steel up to about $\frac{1}{8}$ in. thick. Work commences at the right hand end of the seam and moves towards the left-hand end, the welding rod preceding the blowpipe which moves along with its flame pointing in the direction of travel. The blowpipe is given a small and continuous sideways movement, while the welding rod is moved steadily without sideways movement.

The filling metal is added by a backward and forward piston-like movement of the rod, the rise and fall of the rod allowing the heat to melt the bottom edges of the joint just ahead of the welding pool. The rod must not be held continually in contact with the molten metal, as this prevents the heat reaching the lower edges of the joint.

Rightward Welding

In Rightward welding, the operation is carried out from left to right, the blowpipe preceding the welding rod along the seam. The movements of the blowpipe and the welding rod are quite different from those used when the Leftward method of welding is adopted.

The blowpipe points in the direction of the completed weld, and moves steadily along the seam without lateral movement. The welding rod, however, is directed away from the completed weld and it given a circular progressive movement. The cone of the flame is directed towards the bottom of the seam so that the maximum amount of heat may be utilised.

Rightward welding is quicker and cheaper than the Leftward method on plate thicker than $\frac{3}{16}$ in. With the Rightward method, no bevel is necessary on plate up to $\frac{1}{4}$ in. thick and a 60 degree bevel is sufficient above this thickness. On thicker sheets, the amount of welding rod used in Rightward welding is only about 68 per cent of that required for Leftward welding.

There are other advantages in using the Rightward method of welding. When welding by the Leftward method, it is impossible to get a clear uninterrupted view of the bottom edges of the seam without removing the end of the welding rod. In practice the operator has to move his welding rod from time to time, with the result that not only is the speed of welding reduced,but the hot end of the rod becomes exposed to the atmosphere, and is quickly covered by a layer of oxide scale. The rod is then plunged into the molten metal, depositing flakes of scale into the weld. This causes uneven welds, and a generally unsatisfactory job.

In the case of Rightward welding, the operator's eye is in a direct line with the bottom of the weld, and it is unnecessary to remove the welding rod or interrupt the work to see whether welding is progressing satisfactorily.

There is generally less distortion on work which is welded by the Rightward technique; this is because in this method the temperature of the plate at a given distance away from the seam is lower, the heat being confined more nearly to the seam itself.

A further advantage in Rightward welding is the annealing effect as the flame from the blowpipe is directed over the completed weld. If a metal is heated up to melting point and then allowed to cool suddenly, the resulting structure is coarse, the metal brittle, and possessed of poor mechanical properties. If, however, the metal is allowed to cool down more gradually, as occurs in Rightward welding due to the location of the flame, much better mechanical properties are obtained. There is also the protective effect. Metal that has been heated to a high temperature very rapidly becomes oxidised by the oxygen in the atmosphere, and unless some precautions are taken to keep the atmosphere away from the hot metal during the early stages of cooling, not only oxidisation, but also occlusion of nitrogen will take place, resulting in the production of brittle welds. In the case of Rightward welding, the completed weld is covered by the envelope of the blowpipe flame during the critical cooling period, and the better appearance and properties of welds carried out by this method are to a large extent due to this feature.

The All-position Rightward Method

The all-position Rightward technique is a modification of the Rightward method of welding, whereby the blowpipe flame precedes the rod. This technique is a good one for the welding of mild steel plate and pipe in the vertical and overhead positions. It enables the operator to attain a uniform bead of penetration and an even build-up or reinforcing bead. The welder can work with complete freedom of movement and has a clear view of the weld pool and the fusion zone of the joint.

Vertical Welding

By arranging the welding seam in the vertical plane, certain advantages are obtained. Steel plates from $\frac{1}{16}$ in. to $\frac{5}{8}$ in. thick can be welded without bevelling the edges, though for material over $\frac{3}{16}$ in. thickness, two operators are desirable.

Welding is started at the bottom of the seam and carried out in an upward direction. The welding rod is fed into the weld pool in front of the blowpipe, preceding the blowpipe along the seam. In starting to weld, the edges of the work should be heated sufficiently to fuse them through for their full depth, producing a small round hole between the edges; this hole must be maintained as welding proceeds.

For welding $\frac{1}{16}$ in. thick steel sheet, the blowpipe should be held at an angle of about 30 degrees to the work; for $\frac{1}{8}$ in. material, this angle should be about 60 deg., and for $\frac{3}{16}$ in. sheet about 80 deg.

The Welding of Copper

Copper is a metal that does not readily lend itself to fusion welding; this is because of its great affinity for oxygen, when heated. However a grade of copper has been introduced known as deoxidised copper which overcomes this difficulty. In the final stages of manufacture of deoxidised copper, a small amount of deoxidising element, such as phosphorus, is added to the melt. The phosphorus unites with the oxygen in the copper to form a volatile oxide of phosphorus, which evaporates leaving the metal free from oxygen. At the same time, a small quantity of phosphorus is left in the metal to act as a deoxidiser during welding operations.

The correct welding rod and flux should be used (see Table VI), the rod being thicker than that used in welding steel of equivalent thickness. The welding nozzle should be about two sizes larger than that used for steel. A neutral flame is used, and the Leftward method of welding. The material is preheated in the neighbourhood of the weld to ensure adequate fusion of the edges, and not just adhesion of the deposited metal. The vertical method of welding also gives good results on copper.

Where possible, the weld may be hammered while still hot, and before the temperature has fallen below 600 deg. C., as this improves the mechanical properties of the deposit, consolidates the surfaces and removes porosity.

Although fushion welding can be employed on copper tubes, this process cannot be recommended in boiler work, brazing or silver-soldering being preferred.

Welding Stainless Steel

Only certain types of stainless steels can be welded safely; careful enquiries should therefore be made with the metal suppliers before proceeding.

Stainless steel is welded in a very similar way to mild steel, except that the appropriate flux must be used, and the correct welding rod according to the type of alloy. The flux may be applied to the underneath surface of the metal along the seam. The flame must be strictly neutral and the welding rod kept in the flame the whole time that welding is in progress. Welding should be carried out as quickly as possible, and on completion, the flame may with advantage be withdrawn from the weld slowly, to avoid cracking.

After welding, all oxide and scale should be removed, and the weld can then be ground, buffed or polished as required.

Flame Cutting

The cutting of iron and steel by the oxygen process is carried out by the direction of a closely regulated jet of pure oxygen on to an area previously heated to ignition temperature, and as the iron is oxidised, moving the oxygen jet so that a narrow cut is formed. As only the metal in the direct path of the oxygen is acted upon, very accurate results can be obtained.

The cutting blowpipe comprises a central nozzle supplying the cutting jet of oxygen surrounded by an annulus through which pass the fuel gases for preheating. For hand cutting, the preheating flame is set to neutral and the blowpipe rested on its guide so that the white cone is about $\frac{1}{8}$ in. from the surface of the work. Cutting should be started at the edge furthest away from the operator; when the metal reaches a bright red colour, the cutting oxygen valve is opened, releasing the oxygen stream. The blowpipe is then drawn slowly towards the body along the line of cut. Roller guides are available, to assist in keeping the cut straight, also circle cutting attachments.

Flame cutting is most useful in boiler work, especially when mild steel formers are required for flanging the various plates.

Safety in the Use of Oxygen and Acetylene

It is most important that oxygen and acetylene cylinders, or in fact any cylinders containing gas at high pressure, are treated with care.

Gas cylinders must never be tampered with in any way; their colour, valve threads and markings must never be altered. The user should refuse to accept delivery of any gas cylinder of which the contents are not clearly indicated by labelling, marking or colour. Gases must never be mixed, nor must any attempt be made to fill one cylinder from another.

It is also very important that gas cylinders are properly stored when not in use. Store rooms should be well ventilated top and bottom. Oxygen cylinders must be kept well away from Acetylene cylinders, preferably in a different store room. If cylinders have to stand in the open for a time, they must be protected from ice and snow in cold weather, and from the direct rays of the sun in hot weather. Cylinders should never be laid on wet soil; they should be kept away from rusting or corroding conditions.

No artificial heat is required in gas store rooms. Lighting may either be of the approved flame-proof type, or be outside the building, so that the interior is lit through fixed windows. Electric switches should be flame-proof or placed outside the store room.

Oxygen cylinders are always stored in a horizontal position, being laid on planks with wedges to prevent them rolling; they may be stacked up to four high if properly wedged.

Acetylene cylinders are stored in an upright position. Full and empty cylinders should be kept apart in the store room.

Gas cylinders must never be used as rollers, work supports or jacks. They should be stored well away from heat of any kind, as if exposed to heat, the cylinder walls may be weakened, and the gas pressure may be increased to a dangerous degree. It is also important to remember that oil or grease will ignite violently in the presence of oxygen, so great care should be taken to keep these substances away from gas cylinders or valves; greasy overalls, gloves, etc., should also be kept away from cylinders.

Gas valves or fittings must never be lubricated, nor must white or red lead or other jointing compound be used in connection with them.

Oxygen has no smell, and whilst it does not burn, it supports and accelerates combustion. Propane and acetylene have unmistakable smells, and will ignite and burn instantly from a spark or even a piece of hot metal. Care should be taken to avoid leakage of gas—this can be tested with soapy water, never with a naked flame.

Acetylene can form explosive compounds in contact with certain metals or alloys, in particular alloys of copper and silver; acetylene should therefore never be allowed to come into contact with copper fittings or pipes or any alloy containing more than 70 per cent of copper.

If an acetylene cylinder should become heated accidentally, it must be dealt with promptly as follows: shut the valve, detach the regulator and take the cylinder outdoors into open air immediately. Apply cold water copiously to cool the cylinder, open the valve fully and keep cool with water until the cylinder is empty. As this may take some time, the manufacturers should be contacted for advice.

No welding or cutting apparatus should be used without the proper automatic pressure regulators. Before putting a regulator on to a full cylinder, the adjusting screw for regulating the pressure of output should be released, otherwise there is a risk of damage to the regulator.

Care must always be taken of the hoses used in oxy-acetylene work; they should be inspected periodically to ensure that they are free from cuts, cracks, burns and worn places, and arranged so that they cannot be cut by contact with sharp edges or corners, falling metal, sparks, or the flame from the blowpipe itself. It is advisable to have the hoses of equal length.

When welding or flame cutting, the approved type of goggles must be worn. Ample ventilation should be constantly maintained, and great care should be taken to ensure that all work involving gases under pressure is carried out in a safe place, away from the risk of fire. In any case, a fire extinguisher and both sand and water should always be readily available in case of emergency.

CHAPTER FOUR

Silver Soldering and Brazing

SOLDERING and Brazing are processes which involve the joining of metals by the addition of molten filler metal of substantially different composition, at temperatures well below their melting point. A second important difference between soldering or brazing and welding is that the brazing alloy or filler metal must be drawn by capillary forces right through the mating joint surfaces, and not merely deposited at the edges.

Soft solders are generally accepted as including the alloys of low melting point up to about 400 deg. C. They are usually based on tin or lead with small additions of antimony and sometimes silver.

Hard solders, or silver solders are those alloys suitable for soldering which have melting points from 400 deg. C. up to about 800 deg. C.

Brazing alloys are alloys suitable for brazing which have melting points from about 800 deg. C. up to about 1,000 deg. C.

It should however be understood that there is no definite line of demarcation between silver solders and brazing alloys. The former always contain a proportion of silver, while the brazing alloys generally contain copper, zinc, and sometimes tin. Brazing alloys containing copper and phosphorus are also available, but these are not recommended for boiler work.

Some silver solders and brazing alloys, when used on copper, require no flux. Examples of these are "Silfos" and "Silbralloy", manufactured by Messrs Johnson Matthey & Co., and "Sifcupron", manufactured by Suffolk Iron Foundries Ltd.

Table VII gives details of the silver solders and brazing alloys which the Author has found very satisfactory for model boiler work, and gives the type of flux recommended in each case.

Referring to Table V, the first seven alloys listed are normally used in copper and mild steel boiler work, and the first, Easy-flo, which contains the most silver, is the most expensive. This alloy has very high ductility and very good corrosion resistance. Easy-flo melts very suddenly when used on a job heated by a torch, forming small but neat and smooth fillets. As its melting range is so short, it is an excellent silver solder for use in joining copper parts of thin section, where there is always a danger of burning the metal. Easy-flo is therefore ideal for such work as fixing thin-gauge boiler tubes to tubeplates, and for silver soldering small fittings made of brass or gunmetal.

TABLE VII: PROPERTIES OF JMC SILVER SOLDERS AND BRAZING ALLOYS

Alloy	*Melting range deg. C.*	*Ultimate tensile strength, tons per sq. in.*	*Vickers hardness*	*Flux required*
Easy-flo	620–630	30	131	Easy-flo
Easy-flo No. 2	608–617	30	135	Easy-flo
Argo-flo	605–651	32	136	Easy-flo *
Argo-swift	607–700	30	140	Easy-flo *
Argo-bond	616–735	30	148	Easy-flo *
C.4	740–780			
B.6	790–830	28	121	Tenacity 4.A
Argo-braze 56	600–711			Tenacity 5
Argo-braze 59	611–746			Tenacity 5
Mattibraze 34	612–668	32	140	Easy-flo

* When using these alloys on stainless steels, Tenacity flux No. 6 should be used.

Easy-flo No. 2 is very similar to Easy-flo, but having less silver in its composition, is slightly cheaper. It is not quite so ductile as Easy-flo, and its corrosion resistance, though very good, is less high. The fillets formed by Easy-flo No. 2 are not quite so smooth and neat, it is however a most useful alloy for boiler work, and can be used on almost any part of a copper or steel boiler; it is also useful for joining dissimilar metals. When used for brazing stainless steels, Tenacity flux No. 5 or 6 should be used for both Easy-flo and Easy-flo No. 2.

Both Easy-flo and Easy-flo No. 2 can be obtained in rods $\frac{3}{16}$ in. diameter down to 16 s.w.g., in cut strips $\frac{3}{16}$ in. by 0·040 in. to $\frac{1}{8}$ in. by 0·025 in., in wire from $\frac{1}{16}$ in. diameter down to 0·003 in. and in rings down to $\frac{1}{8}$ in. diameter by 0·010 in.

For some brazing operations it is desirable to provide a more substantial fillet than is obtained when using Easy-flo or Easy-flo No. 2. On these occasions Argo-flo, Argo-swift or Argo-bond may be used. These alloys contain a smaller proportion of silver than the Easy-flos, and are therefore cheaper. They can be used on almost any metal, and are very suitable for boiler work for such operations as brazing throatplates to barrels, backheads to wrappers, foundation rings, etc. These three alloys can all be obtained in rod form, flat strips or as wires in coil.

C.4 is a most useful brazing alloy as a first step in two stage brazing, when followed by Easy-flo or Easy-flo No. 2. B.6 is also a very useful alloy where a higher melting temperature is required, or as a first step in two or three stage brazing.

Whenever brazing alloys having long melting ranges are used, the work should be heated as quickly as possible to minimise liquation effects, and to save unnecessary wastage of the alloy.

Liquation can be explained as follows: Take the case of an alloy of rather low silver content and long melting range in contact with a capillary passage or gap between two pieces of metal to be brazed together. The work is in the process of being heated; when the temperature reaches the solidus of the brazing alloy, some of the alloy melts and runs into the gap, away from the un-melted portion of the alloy. The un-melted part is now no longer in contact with the liquid that has run away, and it has a different composition from that possessed by the original complete alloy. Consequently the melting temperature of the solid portion robbed of its liquid is higher than the quoted liquidus temperature of the original alloy. As a result, it either remains at the mouth of the joint as a costly and possibly unsightly fillet, or the temperature has to be raised much higher than it was originally.

Fluxes for Silver Soldering and Brazing

The manufacture of fluxes for silver soldering and brazing has become a highly specialised business, and the boiler maker will be well advised to purchase his fluxes ready made.

It is important that the correct flux is chosen for a particular silver solder or brazing alloy and for the metals being joined. The Easy-flo flux (Green label) is the standard flux suitable for the majority of the metals in common use, such as brass, copper, mild steel etc. It is active at temperatures below those at which silver brazing alloys melt, has a satisfactory life at brazing temperatures and is an effective solvent for most metallic oxides.

Easy-flo flux is of the fluoride type and should not be employed for brazing operations conducted much above 750 deg. C. The flux residues are soluble in water, the degree of solubility decreasing the longer the flux is kept at high temperatures.

Tenacity flux No. 5 (Red label) is a flux having a higher initial melting temperature than Easy-flo flux and the ability to perform satisfactorily at high temperatures and during protracted brazing operations. This flux can be recommended for brazing steel boilers and particularly for stainless steel boilers or large assemblies in mild or stainless steel. It can also be used on copper-base alloys in positions where only a small amount of flux can gain access to the joint area.

The residues of Tenacity flux No. 5 are only slowly soluble in water, and the solubility decreases as the time and temperature of the brazing operation increase. The most effective solvent, although rather slow in action, is a hot solution of caustic soda. This must of course be very carefully washed off before proceeding with the job.

Tenacity flux No. 4A (Yellow label) is a compromise between an active flux of the fluoride type and a relatively inactive flux of the borax type. It is suitable for use on all varieties of metal where brazing temperatures higher than 750 deg. C. and up to about 1,000 deg. C. may be involved. Its residues

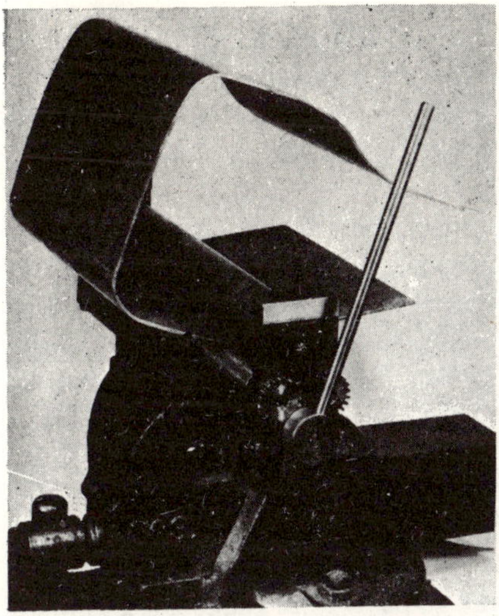

The bending rolls in use on a firebox wrapper

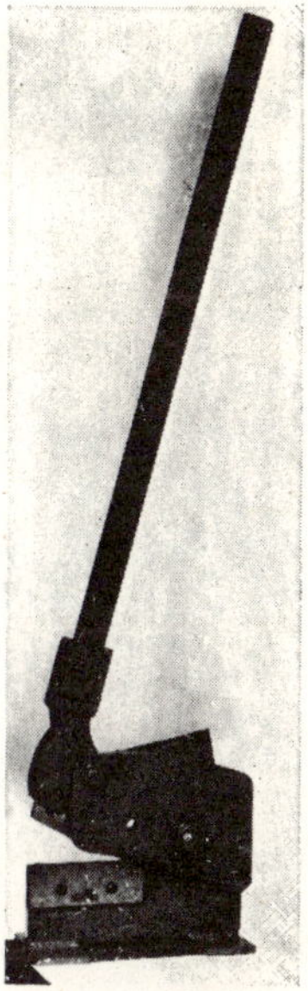

A useful shear for model boiler work

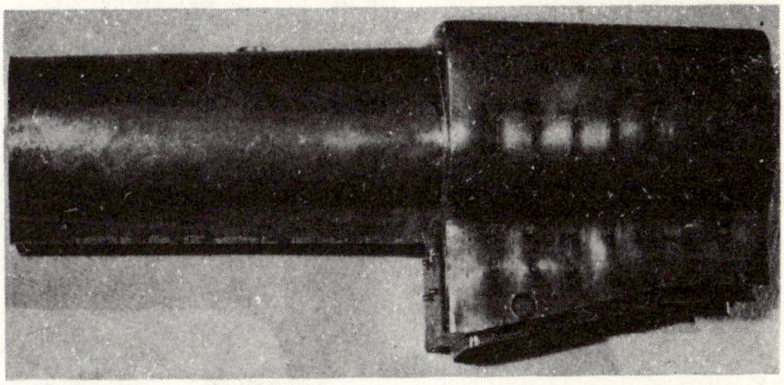

Right: A $3\frac{1}{2}$ in. gauge Belpaire boiler under construction by A. F. Farmer

Below: An example of the author's popular "Rob Roy" design, built by Bertie Green

Flanged plates, syphon and crownstays for the author's 3½ in. gauge 2–6–4 T. made by A. F. Farmer

Right: plates for a "Simplex" boiler made by G. H. Corlett of Transvaal

Bottom left: Top of firebox showing syphon

Bottom right: side view of Belpaire firebox

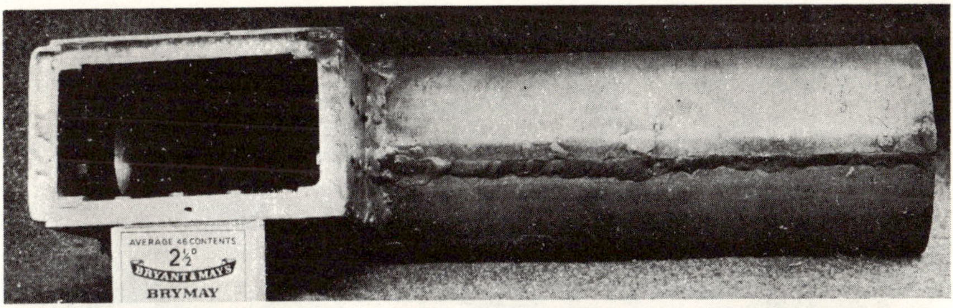

3½ in. gauge boiler with sif-bronzed barrel joint by G. M. Cashmore

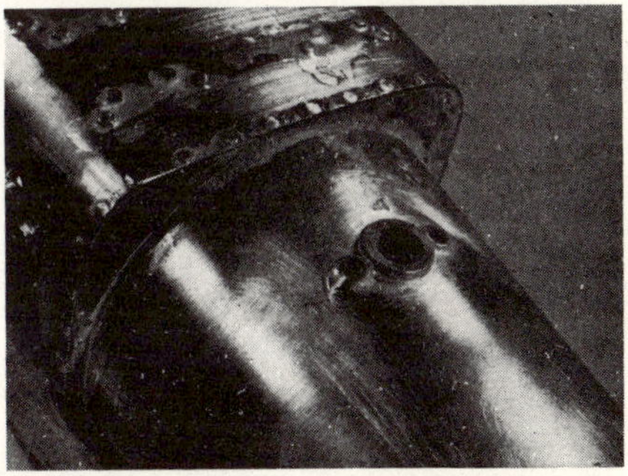

Bushes for safety valve and top feeds on G.W.R. boiler.
Below: A flanged throatplate for a 7¼ in gauge boiler

Underside of 3½ in. guage Belpaire boiler showing syphon; made by A. F. Farmer

Some $\frac{1}{4}$ in. scale boiler plates made by the author

Right: Firebox with crown stays for a 5 in. gauge Belpaire boiler

Below: The same firebox filled wtth tubes and flues

are slowly soluble in water, but warm dilute sulphuric acid is a more effective solvent where the work consists of copper or copper alloys.

Tenacity flux No. 6 (Orange label) can be used when a low-melting brazing alloy is adopted for heat-resisting materials forming refractory oxides, under conditions which are not condusive to good fluxing, i.e. when less than the desirable amount of flux can gain access to the joint faces. It can also be recommended for cases where the maximum strength of bond is desired. This flux is effective at somewhat higher temperatures and for longer times of heating than is Easy-flo flux. It does however suffer from the disadvantage in hand operations of covering the joint with an opaque film through which it is not easy to observe the flow of the brazing material.

After Tenacity flux No. 6 has been mixed to a paste with water, it must not be stored in this condition in a sealed container, or the container may burst.

The Use of Flux

All the fluxes mentioned previously should be mixed with clean cold water to the consistency of cream. All types except No. 5 and No. 5A react with water to a slight degree, and may therefore thicken somewhat during the first ten minutes after mixing, a further small addition of water may then be made. The paste tends to harden due to evaporation if left exposed to the air for long periods, but may be reconstituted for further use by the addition of more water. Should the paste become dirty or dusty in any way, it should be discarded.

Before any joint to be brazed or silver soldered is fluxed, it must be absolutely clean, and slightly roughened with coarse emery cloth. The flux is then applied liberally with a brush. When parts such as boiler flanged plates are to be joined to their respective wrappers, it is often a good plan to put a thin coating of flux all around the join before riveting the parts together; when the job is ready for brazing, further flux is applied along the outside of the seam.

Brazed and silver soldered joints do not always turn out to be free from defects such as flux inclusions and voids. Apart from the more obvious causes such as insufficient brazing material or flux, or under or over-heating, these defects arise mainly from uneven heating, non-uniform joint clearances, or variations in the condition of the surfaces over which the alloy flows.

Torch or furnace heated joints tend to be heated unevenly because the outside surfaces or edges on which the flame impinges, or on which the furnace walls radiate, reach a higher temperature, and more quickly, than the less accessible parts of the joint. If joint edges reach brazing temperature ahead of the rest of the joint, and if the brazing alloy is applied before the whole has reached the right temperature, the alloy flushes round the edges and tends to seal flux and gas inside. Sealed inclusions thus produced do not usually entirely disappear as a result of longer heating or the application of more brazing alloy; the best chance of eliminating them is by scratching the affected area with a steel rod sharpened at the end, and applying further alloy previously dipped in dry flux.

Flux which has become saturated with oxide because of prolonged heating, or through over-heating, is more likely to be trapped; thus quick but even heating of the whole of the joint area is most beneficial. Whenever further additions of flux are required during the heating process, these are better applied dry than as paste. In cases where the joint is being fed by hand using a brazing rod or strip, this can be heated and then plunged into dry flux powder to pick up additional flux for transfer to the joint.

Many low temperature brazing fluxes contain fluoride compounds, they should not therefore be allowed to come into contact with the skin. If the boiler maker accidentally allows the flux to get on to his hands, this is not a matter for serious concern, but he should wash his hands thoroughly as soon as possible after brazing is completed. Any cuts or breaks in the skin must be covered by a proper dressing, as wounds will be delayed in healing or may even become septic if flux is introduced into them.

All fluxes produce fumes when heated, particularly if heated above the temperatures described as their recommended maximum temperatures of use; adequate ventilation should therefore always be provided wherever brazing operations are taking place.

Heating Methods for Brazing

There are several heating methods available for silver soldering and brazing; they include the following: Torch brazing, Furnace brazing, High frequency induction heating, Carbon resistance brazing, Direct resistance brazing, Dip brazing and Salt bath brazing.

Torch brazing is probably the only method really suitable for small boiler work, and the following are in general use: Paraffin (Kerosene) blowpipe, Petrol (Gasoline) blowpipe, Air-Coal gas burner, Oxygen-Coal gas blowpipe, Oxygen-Hydrogen blowpipe and Oxy-Acetylene blowpipe.

Paraffin Blowlamps

Paraffin (Kerosene) blowlamps provide a suitable source of heat for model boiler work up to the smaller $\frac{3}{4}$ in. scale locomotive boilers. They can be obtained in sizes from about $\frac{1}{4}$ pint up to about 7 pints, though blowlamps below about 1 pint are too small for most model locomotive boilers.

The well-known "Sievert" paraffin brazing lamps are made in ten sizes, from $1\frac{1}{2}$ pints to 7 pints. The $1\frac{1}{2}$ pint size, No. 569, has a container which normally takes $1\frac{1}{4}$ pints of paraffin, the burner tube is $\frac{7}{8}$ in. diameter, and the length of flame produced at normal pressure about 9 in. Average consumption is $1\frac{1}{4}$ pints per hour, and maximum consumption about 2 pints per hour. Such a blowlamp will provide enough heat to silver solder a small Gauge I locomotive boiler.

Coming to the larger sizes of paraffin brazing lamps, the Sievert No. 577 takes about 4¾ pints of fuel and consumes from 2⅝ pints to 5¾ pints per hour, giving a flame about 12¼ in. long. The burner tube is 1⅜ in. diameter, and the lamp weighs about 8½ lb. No. 579 takes about 6½ pints of paraffin and consumes from 3¾ pints to 7¾ pints per hour. The flame from this brazing lamp is about 12¾ in. long, the burner tube is 1⅞ in. diameter, and the weight of the lamp complete about 11½ lb.

No. 577 brazing lamp may be regarded as capable of brazing all types of locomotive boilers up to ½ in. scale and the smaller ¾ in. scale boilers. No. 579 will braze most ¾ in. scale boilers. With the assistance of an additional blowlamp of about 1½ pint size, both No. 577 and No. 579 would cope with practically any size of boiler in 1 in. scale.

The brazing lamps referred to above are all fitted with self-contained pumps and can be obtained with a quick starting device, whereby fuel from the tank is utilised to start the burner, eliminating the use of methylated spirit or alcohol (or other dirtier methods) for pre-heating. This device can be recommended as it saves considerable time in starting up.

For the purpose of comparison with other forms of heating, it may be said that the paraffin brazing lamp No. 579 can give a heat output of about 142,000 British Thermal Units per hour. The highest temperature reached in the flame is about 1,600 deg. C.

Petrol Blowlamps

Petrol (Gasoline) blowlamps give a very similar performance to paraffin blowlamps, though the flame produced is longer and gives slightly more heat.

The Sievert range of petrol brazing lamps are obtainable in sizes from 1½ pints to 7 pints. The Sievert lamp No. 268 takes 4¾ pints of fuel and gives a flame approximately 22 in. long from a burner tube 1⅝ in. diameter. Consumption of this lamp is about 5⅝ pints per hour at normal pressure, while maximum consumption is about 11 pints per hour. The Sievert lamp No. 269 is similar to No. 268 but has a larger fuel container taking 6½ pints of petrol. Both No. 268 and No. 269 petrol brazing lamps give a maximum heat output of about 187,000 British Thermal Units per hour. The maximum temperature reached in the flame is about 1,600 deg. C.

Air-Coal Gas Blowpipes

Air-coal gas blowpipes are very useful appliances for silver soldering and brazing model boilers. The smaller sizes of blowpipe can be supplied with air from a foot blower, but a rotary blower driven by an electric motor is to be preferred.

For model boiler work, an air-gas blowpipe having separate taps for both the air supply and the gas are to be preferred. The "Radiation" 8C. pattern blowpipe can be obtained in four sizes, the largest (No. 4) has a gas inlet diameter of 1⅛ in., and an air inlet diameter of ¾ in. Provided with a gas supply of 1 in. nominal size and a suitable air blower, this blow pipe is capable of brazing locomotive boilers of ¾ in. scale and the smaller 1 in. scale boilers without assistance. With assistance from an additional blowlamp or Propane torch, boilers of almost any size can be tackled.

The Radiation No. 3 air-gas blowpipe has a gas inlet diameter of ¾ in., and an air inlet diameter of ⅝ in., and is designed for use with a ¾ in. gas supply. Provided with a suitable air blower, this blowpipe will cope with most model locomotive boilers up to ¾ in. scale.

It is important to remember that an air-gas blowpipe will not function properly unless the gas supply all the way from the meter is large enough for the blowpipe in use. In cases of doubt, the local Gas Company should be consulted. To prevent any possibility of a "blow-back", a light non-return valve must always be fitted on the gas supply to the burner.

Provided that both the gas and the air supply are sufficient, and the burner properly adjusted, these air-gas blowpipes will give a flame temperature (in the hottest part of the flame) as high as 1,750 deg. C.

One advantage of the air-coal gas blowpipe over other types of burners suitable for brazing work is the fact that the gas used is considerably cheaper than such gases as acetylene, propane or butane. There is also the point that a steady and continuous supply of gas is always available.

Suitable air blowers can be obtained from Radiation Ltd, and from the larger dealers in model engineers' requisites; in most cases, it will be found that a standard ¼ h.p. electric motor running at about 1,450 r.p.m. will drive the blower satisfactorily, the blower running at around two thirds motor speed.

Propane Blowpipes

Burners using bottled gas are extremely convenient for all heating purposes, including of course silver soldering and brazing. Bottled gas may either be Propane (C_3H_8) or Butane (C_4H_{10}), mixtures of these are also available. The fuel is stored in liquified form under pressure in steel containers, and vaporises when the regulating valve of the container is opened. For both workshop use and outdoor work propane is to be preferred; it is always used with a pressure-reducing valve or a hose-failure valve.

Propane is non-poisonous and burns with a clean flame, producing no soot or smell. To serve as a warning in case of leakage however, the gas has been given a characteristic smell. It is heavier than air and has very narrow explosion limits. The effective calorific value is about 20,000 British Thermal Units per pound. The theoretical combustion temperature with air is about 1,900 deg. C. (3,450 deg. F.).

Propane is obtained in containers provided by the gas distributors, who also supply full containers in exchange for empties. For small jobs, the Sievert small gas containers are recommended. The gas pressure in a propane container is normally about 100–140 lb. per sq. in. The Sievert gas burners should however work on a gas pressure of 14–56 lb. per sq. in., so that the container pressure must be reduced. For this purpose Sievert supply special reducing valves with or without a built-in hose-failure valve. The hose-failure valve is a safety device which automatically shuts off the gas flow in the event of hose failure.

The Sievert propane burners are made in a very wide range to suit all kinds of work. No. 3938 is ideal for very small silver soldering, boiler fittings, etc., It weighs only $1\frac{3}{4}$ oz., and consumes $\frac{3}{4}$ oz. of gas (at average pressure) per hour. Sievert No. 3939 is slightly larger than No. 3938 and consumes $2\frac{1}{4}$ oz. of gas per hour.

For boiler silver soldering and brazing, Nos. 2942, 2943 and 2944 are suitable. Sievert No. 2942 weighs $4\frac{1}{4}$ oz., consumes 42 oz. of fuel per hour, and will silver solder locomotive boilers under $\frac{1}{2}$ in. scale. No. 2943 weighs $4\frac{1}{2}$ oz., consumes 70 oz. of gas per hour, and will deal with boilers of $\frac{3}{4}$ in. scale. Burner No. 2944 gives a very large and powerful flame, consuming 145 oz. of gas per hour and giving a maximum heat output of approximately 250,000 British Thermal Units per hour. The highest temperature reached in the flame is around 1,900 deg. C. No. 2944 will silver solder most boilers up to one inch scale.

For model boiler work, the most suitable burner support is Sievert No. D.3. This is fitted with brass handles with a bakelite hand grip, an iron shackle and a suspension hook. It is also provided with an adjustable economiser valve. While in use the operator's grip on the handle holds the valve open and the flame burns at full strength. When the finger releases the trigger of the valve, the flame automatically reverts to a pilot size. Vertical supports and horizontal supports with adjustable stirrups are also available.

The Sievert small gas containers are available in three sizes: $12\frac{3}{4}$ oz., $2\frac{1}{4}$ lb., and $4\frac{3}{8}$ lb. The largest size will supply burners up to No. 3941 for a reasonable period, but for the brazing of boilers where the larger burners are used, a larger container should be obtained from the gas distributors.

It is most important that the correct high-pressure hose, as supplied by the manufacturers, should be used for connecting the burner support to the container. A gas pressure regulator, preferably with built in hose-failure valve, must be fitted to the container.

Propane burners are extremely simple to operate and require no pre-heating. As even burner No. 2944, when fitted to the D.3. support, only weighs just under 2 lb., and as there is only one hose connected to the support, it will be seen that the system is a very easy one to handle.

Oxy-Acetylene Blowpipes

In addition to its use for welding, the oxy-acetylene blowpipe can also be used for silver soldering and brazing work; there is however a danger of over-heating and burning the work, especially thin copper sections, boiler tubes, etc. The beginner is therefore advised to use other heating methods in boiler making until experience has been gained. In the hands of the expert, the oxy-acetylene blowpipe can do practically any heating job connected with boiler work, and unlike other burners or blowlamps, is not limited to work of comparatively small size and weight.

When using oxy-acetylene for silver soldering, it is most important that the flame is kept on the move all the time, and not allowed to play direct on the silver solder or brazing alloy; only the outer part of the flame should be used, and the flame should be a slightly oxidising one. It is advisable to coat the joint inside and out with the appropriate flux before heating; as soon as the joint reaches the required heat, the silver solder or brazing strip is dipped into dry flux and quickly applied to the joint. The alloy should then run cleanly along the joint, forming a neat fillet, but if it does not run freely, the alloy is removed and the flame applied again, keeping the flame moving all the time. With practice it will be found that the quicker the job can be done, the neater the joint produced.

The preparation of the oxy-acetylene equipment for silver soldering and brazing is exactly as described for welding (see the previous chapter) though it should be noted that the blowpipe nozzle chosen should be about two sizes larger than that recommended for welding of the equivalent thickness of mild steel.

Oxy-Coal Gas Blowpipe

The oxygen-coal gas blowpipe is another type of burner which can be considered for silver soldering and brazing. The British Oxygen Company supply a suitable blowpipe, which is available with seven different sizes of nozzle. It is of the injector pattern with valves for controlling oxygen and coal gas. The temperature reached in the hottest part of the flame from this blowpipe is approximately 2,000 deg. C., that is to say, slightly higher than that reached by the propane burners.

Summary of Burners for Silver Soldering and Brazing

The type of burner chosen by the amateur boiler maker will obviously depend on many different factors, such as whether a coal gas supply is available, whether the funds available are sufficient for the preferred equipment, and the experience of the person concerned.

If no coal gas is available, the propane torch can be recommended especially where the operator lacks experience, while the skilled model engineer will probably choose the oxy-acetylene equipment. Given a suitable supply of coal gas, the air-coal gas burner with electrically driven blower can give very good service.

The beginner should always seek the assistance of a skilled and experienced model engineer before attempting to braze his first boiler, especially where oxy-acetylene or oxy-coal gas burners are to be used. Help and sound advice can often be obtained from members of the nearest model engineering society (many such societies possess their own oxy-acetylene plant); even when experience has been acquired, an able assistant can be a great help when brazing boilers, especially those larger than $\frac{3}{4}$ in. scale.

CHAPTER FIVE

Workshop Equipment for Boiler Work

THE workshop equipment required for model boiler work, apart from the burners used for silver soldering and brazing, is neither elaborate nor costly, and much of it can be home-made.

Perhaps the most important item of equipment after the actual burners is a suitable brazing hearth or stand. The size and shape of this will of course depend on the size and type of boilers to be brazed. For locomotive boilers up to $\frac{3}{4}$ in. scale, a rectangular hearth about 2 ft. long by 1 ft. wide by 6 in. deep would be suitable, this being built up from black steel sheet about $\frac{1}{16}$ in. thick and supported on strong legs of angle iron so that the floor is about 2 ft. 6 in. from floor level. For locomotive boilers in one inch scale, a suitable size would be 2 ft. 6 in. or 2 ft. 9 in. long, 15 in. wide and about 7 in. deep.

A boiler brazing hearth should have a circular hole about 5 in. diameter (for $\frac{3}{4}$ in. scale boilers, or $6\frac{3}{4}$ in. diameter for one inch scale boilers) cut in the middle of the floor, and arranged so that this can be readily covered by a further steel plate. When brazing the backhead of a boiler, this cover plate can be temporarily removed and the barrel of the boiler put right through, so that the backhead does not lie too high above the hearth for convenience.

Another feature which is very useful but not often found in amateur-made brazing hearths is a swivelling top, the whole upper area being made to swivel about a stout pivot. This enables the operator to direct his flame all round the job, particularly useful when running silver solder over rivet heads, etc. As the making of a swivelling top brazing hearth calls for a great deal more work, the use of swivelling wheels or castors attached to each of the four upright angle irons supporting the hearth is worth consideration. It is however important to ensure that the hearth is properly supported and rests evenly on its legs, for a model locomotive boiler in the larger scales is a very heavy affair, and one cannot afford to take any chances over its safety.

Some brazing hearths are constructed with one end hinged, so that this end can be dropped when it is required to direct the brazing torch upon the end of a boiler. This sometimes occurs while brazing the crown stays of locomotive boilers to the outer firebox wrapper.

In normal use the brazing hearth is filled with coke, while firebricks may be laid around the sides and back of the hearth to reflect the heat back upon the job.

Pickling Tanks

For the silver soldering and brazing of copper boilers, a "pickle" tank is essential. This is a tank containing dilute sulphuric acid in which the boiler or part of the boiler is immersed after each brazing job to remove the burnt flux and clean up the work so that it can be properly examined for possible flaws.

Strong rubber buckets can be obtained which make excellent pickling baths for small boilers, that is locomotive boilers up to the smaller ½ in. scale sizes. For the pickling of larger boilers, a suitable tank can be made by using a strong wooden box and lining the whole of the inside of this with sheet lead. The tank should of course be made long enough to accommodate the longest boiler likely to be built, of width rather more than the overall height of the finished boiler, and as deep as possible, as the greater the depth of the tank the less trouble there will be from splashes.

Another solution to the pickle tank problem is to approach a friendly Builder with a view to obtaining an old kitchen sink. Most "old-fashioned" kitchen sinks have a glazed internal surface, and even though these sinks may no longer find favour with the housewife, they make excellent pickle tanks for the model engineer and boiler maker. A close fitting rubber plug must of course be used, one does not want to risk flooding the workshop floor with sulphuric acid!

The acid itself can be obtained from the local chemist's shop. The commercial grade concentrated sulphuric acid should always be asked for, as this is much cheaper than the pure acid generally supplied to chemical laboratories, and it should be mixed with water in a proportion of about 1 part of acid to 10 parts of water. It is most important to remember that the sulphuric acid must always be added to the water, never the other way about, or a severe explosion may be caused.

Bending Rolls

One of the most useful tools for model boiler work, though seldom seen in the model engineer's workshop, is the bending rolls. These can be obtained commercially, but the smallest size available is likely to prove too heavy for model boilers under 1½ in. scale. However bending rolls are not at all difficult or expensive to make; basically they consist of three rollers, two arranged in fixed bearings and geared together, the third in movable bearings, above and between the other two; by means of pressure screws at each end, the movable roller can be pressed down upon the sheet metal to be bent, which is slid between the three (see Fig. 15).

The length of the rollers should be made slightly longer than the longest boiler barrel or firebox it is expected to roll, and the diameter of the fixed rollers should be about one-twelfth of their length; if made smaller in diameter than this, they may tend to bend in service. Two or three pressure rollers can be made, of different diameters, the smallest being not less than three-quarters the diameter of the fixed rollers, the largest about one and

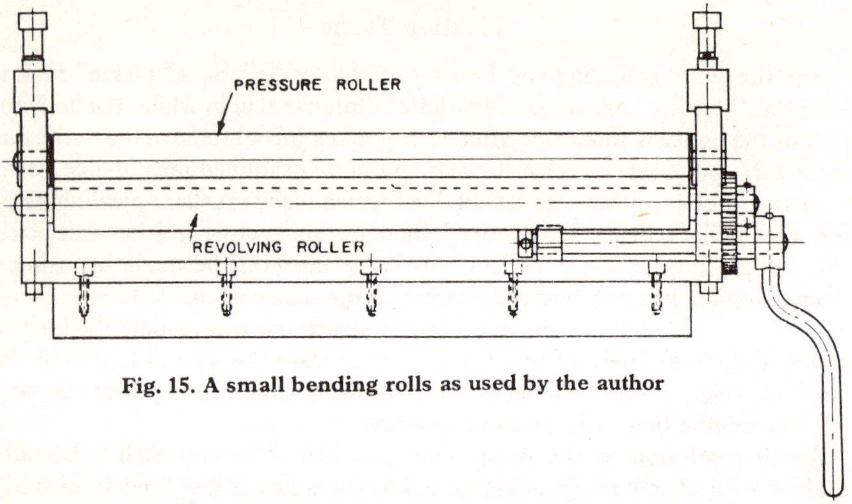

Fig. 15. A small bending rolls as used by the author

one-quarter times the diameter of the fixed rollers. The pressure roller being at the top, and not having any gears attached to it, it can be very quickly taken out and changed for a roller of different diameter. All the rollers should if possible be made of a high-carbon steel, though mild steel will be found satisfactory for work on copper.

In operation, to roll a boiler barrel from flat sheet, the metal is first thoroughly annealed, then placed between the rollers and a little pressure applied to the pressure roller by turning the pressure screws at each end an equal amount. The lever controlling the two lower rollers is then operated, moving the sheet metal from side to side. As the sheet takes on a circular form, further pressure is applied by the pressure screws, the lever is operated again, and so on, until bending is complete, the shape of the sheet metal being checked against something truly circular of the diameter required for the finished barrel, as the work proceeds.

To produce a tapered barrel, the same procedure is followed, but more pressure is applied to one end of the pressure roller than the other, and at this end the resulting barrel will be smaller in diameter than at the other end.

When bends of much smaller radius are required, such as the radiused corners of Belpaire fireboxes, inner firebox wrappers, etc., these can be obtained by locating the sheet metal in the middle of the required bend, and applying an even pressure on the pressure roller without moving the sheet metal at all. In this operation it will be appreciated that the smaller the diameter of the pressure roller, the smaller the radius produced in the work.

Metal Shearing Machines

Another very useful tool for the boiler maker is a metal shearing machine. Although a regulation "guillotine" is of very great assistance for cutting sheet metal without appreciable distortion, this type of machine may be beyond

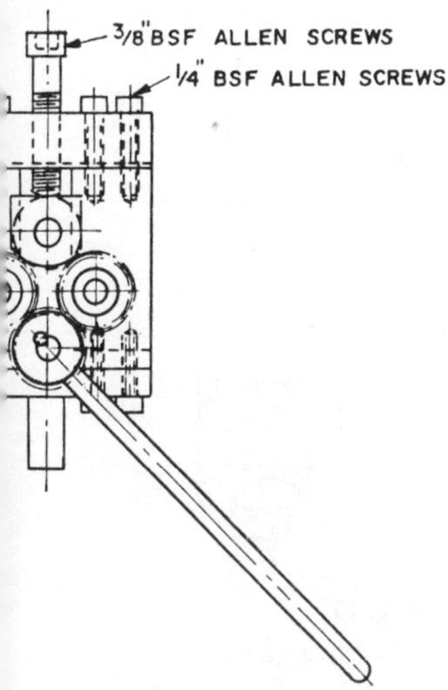

the resources of many model engineers. The ordinary hand lever shearing machine is however quite moderately priced; alternatively a suitable machine can be made up in the amateur workshop without much difficulty (see Fig. 16).

The two cutting blades are made from tool steel or "ground flat stock", and are hardened and tempered after all machining, drilling, etc., has been done. The design of the frame of the shearing machine should be such that the sheet metal being cut can be pushed forward in a continuous motion without fouling the frame. As the pivot pins of both the lever, the connecting links, and the cutting blade are subjected to considerable stresses, they should be made of a high-carbon or nickel-chrome steel.

Riveting Tools

For all types of model boiler work, a selection of rivet punches and "dollies" will be required. Standard single punches and dollies can be obtained ready-made quite cheaply, but to get the best results in boiler riveting, readers are advised to make up their own drawing-up punch and three snap punches to

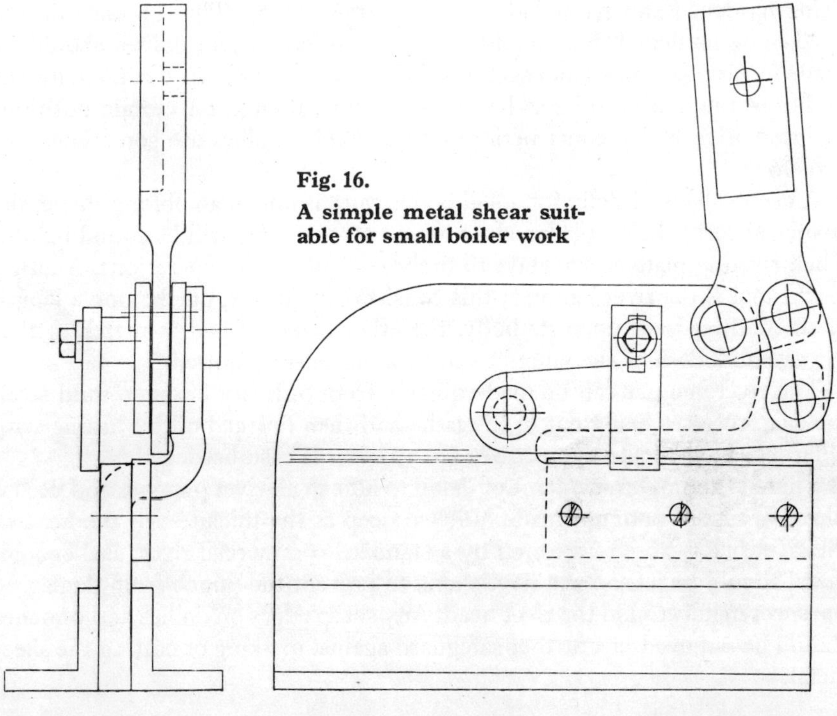

Fig. 16.
A simple metal shear suitable for small boiler work

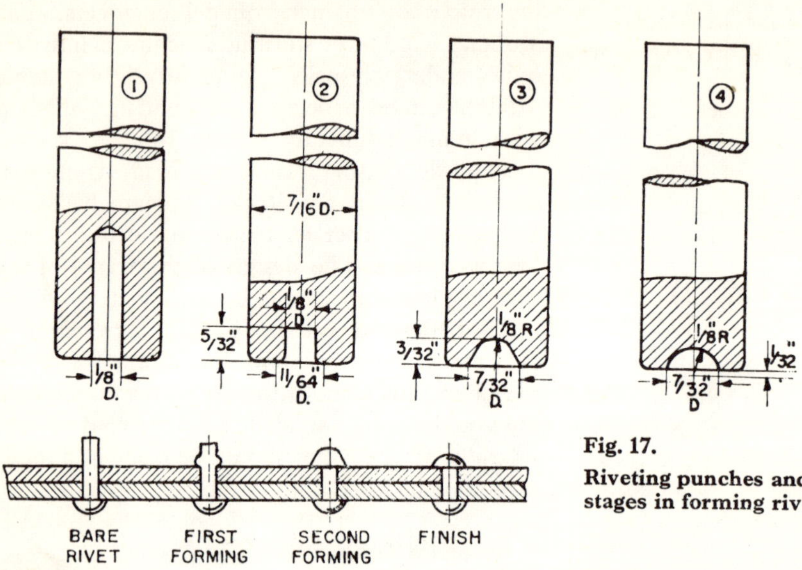

Fig. 17.
Riveting punches and stages in forming rivets

each rivet diameter. Fig. 17 shows what is required; the length of the punches can be around 4 in., though longer punches are sometimes desirable for certain positions.

Silver steel is quite satisfactory for punches and dollies, the business ends being hardened and tempered at a dark straw colour. The top ends do not need to be hardened, but should be well chamfered. The dollies should be made in various lengths and sections. For good riveting, it is desirable for the bodies of the dollies to be as heavy as possible, though for certain positions in locomotive boiler construction, rather slender dollies are sometimes unavoidable.

A useful shape of dolly for small boiler work is one of an oblong shape, the snaphead recess being placed close to one corner. This will be found helpful when riveting plate crown stays to fireboxes and similar operations. Another useful tool is the rivet jammer; this consists of two sets, one having a length of screw thread cut upon its body, the other being tapped to match so that the overall length of the jammer can be adjusted as required.

The rivet jammer can be made quickly from ordinary hexagon mild steel, the snap recesses being cut in the lathe and then finished off by hitting with a hardened steel ball; the recesses are afterwards case-hardened.

There is an important point of detail to note in all rivet punches and dollies the snap recesses are not made quite so deep as the thickness of the head of the rivet being closed, as tested by a standard commercial rivet, and enough metal should be left on the rivet shank to prevent the punch from damaging the sheet metal around the rivet head. Any sharp edges on dollies and punches should be removed as a further safeguard against marking or cutting the sheet metal.

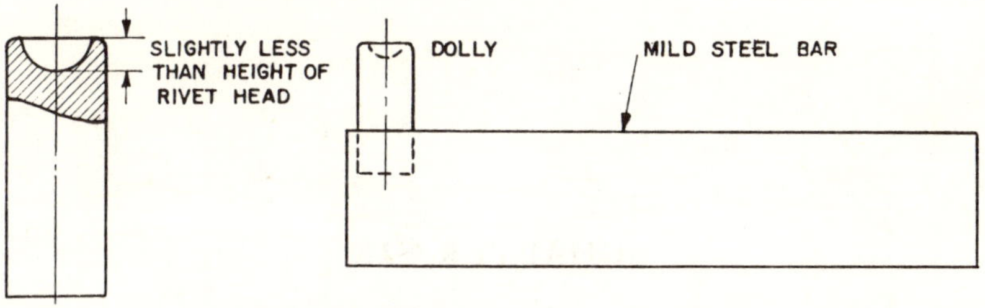

Fig. 18. A useful riveting tool for boiler work

A short riveting dolly and wedge will be found very useful for dealing with firebox crown stays and firebox work generally. While the dolly is hardened, the wedge can be made in mild steel; the angle of taper of the wedge should not be made too great or the dolly may slip while the rivet is being closed; about one deg. is generally satisfactory.

Finally, a few flat-ended punches will prove useful, for bringing plates and flanges together or for use with countersunk rivets which are occasionally used in boiler work, such as when fitting barrels to throat-plates.

Hammers for riveting should be of the ball-pein type, the flat ends of which having slightly rounded edges. Some commercial hammers of doubtful quality will be found to have faces which are not quite square; these should be carefully ground before use. The hammers will vary in weight between $\frac{1}{2}$ lb. and $1\frac{1}{2}$ lb. according to the size of rivet being dealt with. The working faces of hammers should always be kept bright and smooth.

Other tools for boiler work include strong brazing tongs, or tongs suitable for handling the boiler while hot, or for putting the boiler or boiler parts into the pickling tank. These are best made from flat black mild steel strip, the ends being heated to red and twisted through 90 deg; the most suitable shapes for the ends of the tongs will soon become apparent as the boiler maker gains experience.

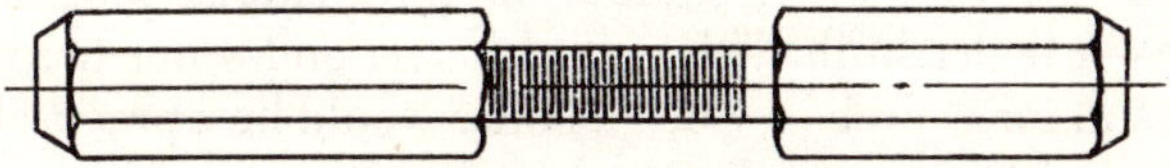

Fig. 19. A rivet jammer

CHAPTER SIX

The Construction of Model Boilers

Choice of Materials

BEFORE the actual construction of the model locomotive boiler is considered, a decision must be made as to the materials to be used.

For all locomotive boilers up to $\frac{3}{4}$ in. scale, an all-copper construction is recommended. Copper is the ideal metal for small boilers, being easy to work, ductile, very resistant to corrosion and easy to solder or braze its only real disadvantage is its high price.

Copper is also an excellent material for 1 in. and $1\frac{1}{2}$ in. scale locomotive boilers, and it is only the high price of this metal which deters builders from using it for the larger model boilers.

Stainless steel has been used with success for model boilers by some of the more experienced workers. It is of course a very strong alloy with good corrosion resistance, but it is hard to work, none too easy to braze or silver solder, and is often difficult to obtain in small quantities. It is not usually stocked by firms specialising in supplies for the model engineer.

The cost of a boiler built entirely from stainless steel is not likely to work out less than one made in copper, so that on balance copper would appear to be the best all-round material, while mild steel will generally be chosen for locomotives of 2 in. scale and above.

There is however one other solution of the material problem worthy of consideration—a boiler built with a steel barrel and outer firebox, but with copper tubes and inner firebox. Such a boiler would be a practical proposition at $1\frac{1}{2}$ in. scale and larger, and for the best results it would be made with all joints silver soldered. It has been suggested that the use of dissimilar metals sets up electrolysis within the boiler, but the effects of this do not seem to be very serious in practice; see Chapter VII.

The Boiler Barrel

Most boiler builders make a start on the barrel. This is perhaps only logical as the barrel is generally the arger part of the whole boiler. If the barrel of the boiler under consideration is a parallel one, a seamless solid-drawn tube should be considered. Although considerably dearer to buy than the equivalent quantity of flat sheet, the seamless tube saves a great deal of time, especially to the less experienced, and the builder is generally assured of a truly circular barrel, making connection to the smokebox an easier operation.

If the design calls for a round-topped firebox, the seamless tube may also be used to form the outer firebox wrapper. The tube is cut cross-wise at the throatplate and a longitudinal cut made from the rear end along the bottom centre-line of the barrel to meet the first cut at right-angles. The two sides are then (after annealing) pulled out and bent to the required shape to form the outer firebox wrapper.

It may of course be found that if the firebox is a deep one, the sides formed in this way are too short to reach to the foundation ring; but the sides can always be extended by silver soldering to their lower edges sufficient sheet material of the same thickness. See Fig. 20. (below).

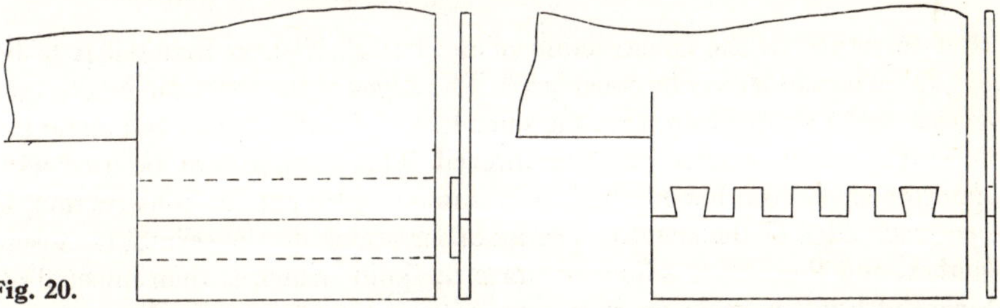

Fig. 20.

The flanged throatplate that is required for a boiler built in this way is a very simple affair, much easier to make than a regulation throatplate (Fig. 21).

Another method of boiler construction is to use seamless tube for the barrel proper, and to bend up the outer firebox wrapper from flat sheet material. The connection is then made by a proper throatplate shaped as in Figure 22 or by a simple throatplate and a turned ring as in Figure 23. This method would normally be used for a Belpaire firebox.

Fig. 21.
A simple throatplate for round-topped fireboxes

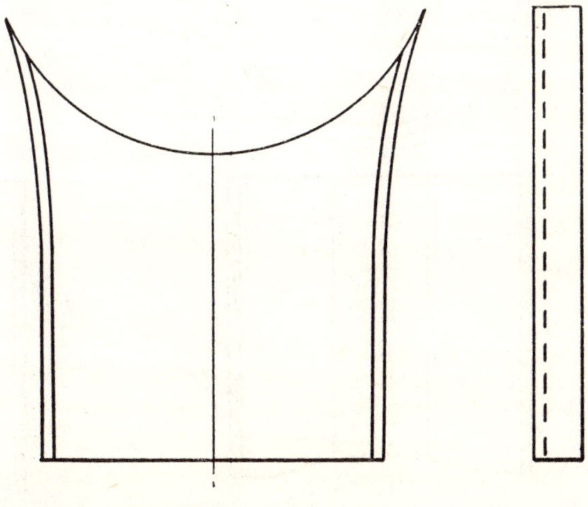

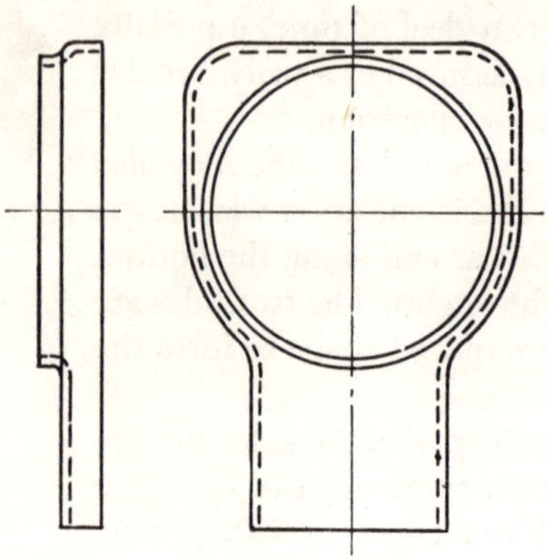

Fig. 22. Throatplate for Belpaire firebox

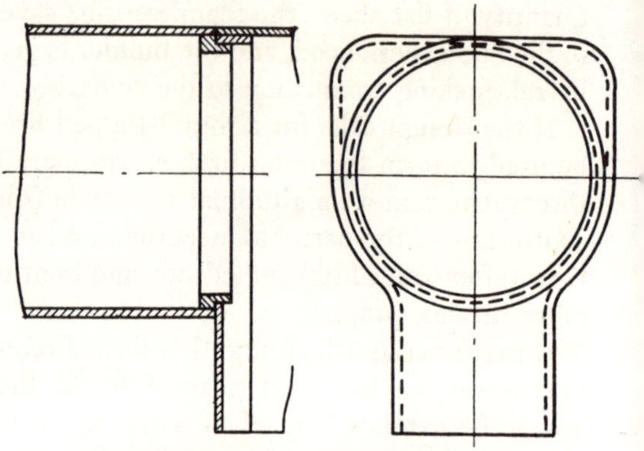

Fig. 23. The use of a turned ring to strengthen a throatplate

But to return to the construction of the barrel, if sheet material is to be used, three methods may be considered. The edges of the sheet can be brought together and a strap made from the same material as the barrel laid along the join on the inside, riveted and then brazed. This method may be used with copper or steel. Another method, only suitable with copper construction, is to cut each edge of the sheet in a series of serrations, fit the serrations closely together, and then silver solder or braze the joint, which is then finished off by careful light hammering over a round mandrel of suitable diameter (see Fig. 24).

Fig. 24. Methods of joining a barrel seam

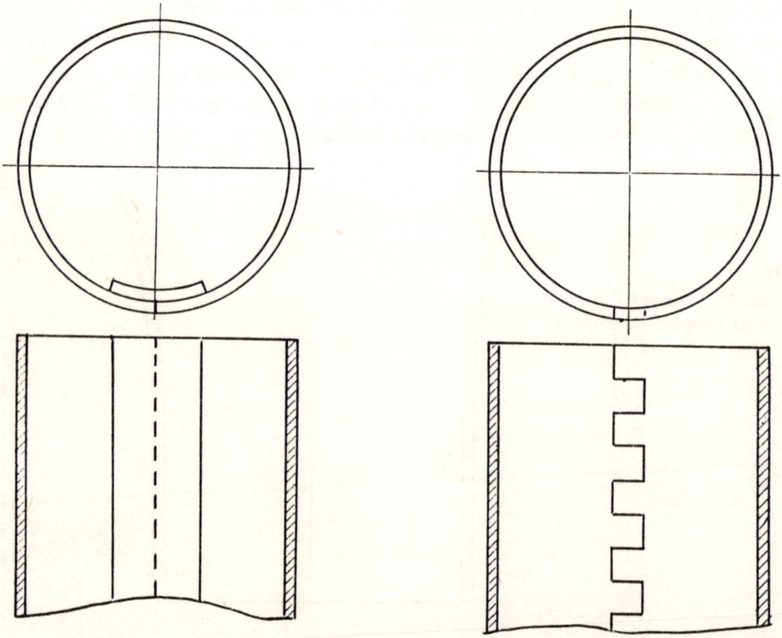

The author's "Firefly" boiler nearing completion

Left: rear view before fitting backhead

Below: the author's $\frac{3}{4}$ in. scale "Jubilee" boiler approaching completion

Firebox backplate with firehole ring fiitted—"Boxhill" boiler

Fire box for "Firefly" ready for the tubes

Right: "Firefly" boiler

Below left: "Firefly" firebox showing inlet of syphon

Below right: The "Jubilee" firebox

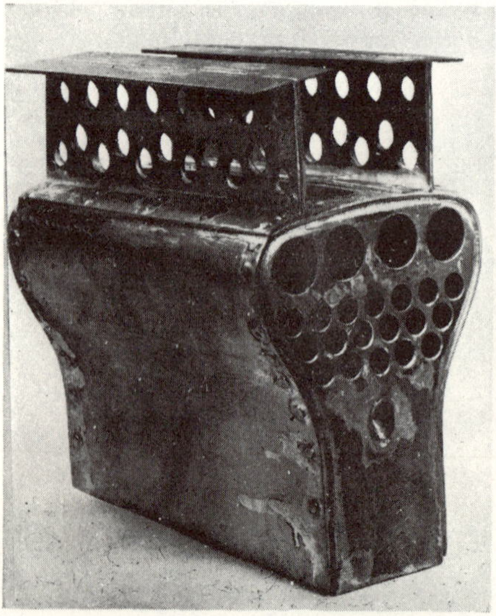

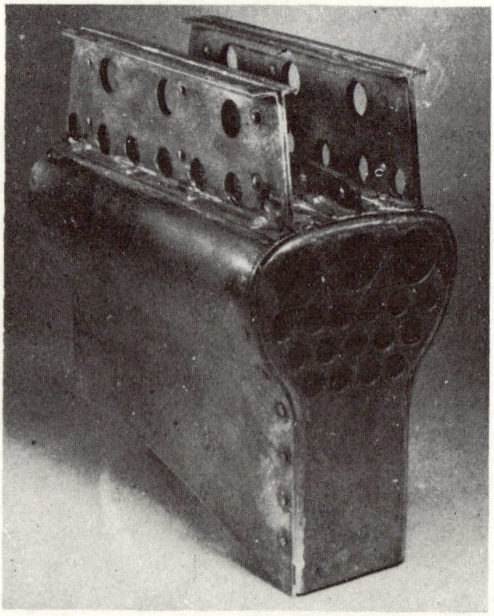

Plate 26

Left: Backhead of Mr. A. F. Farmer's Belpaire boiler

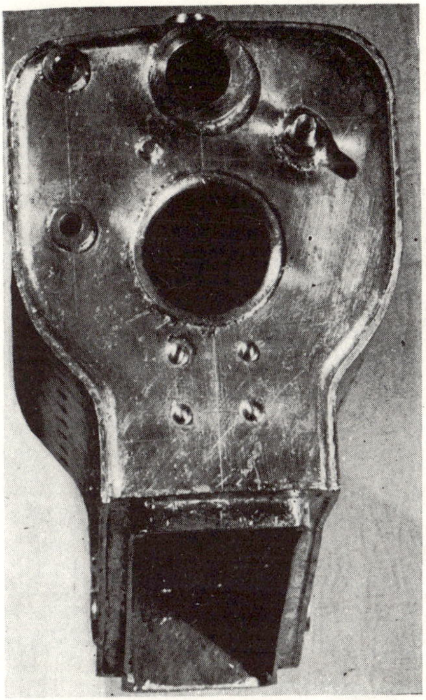

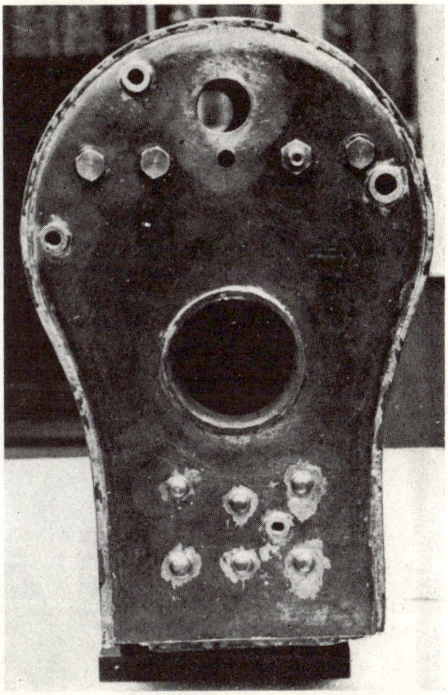

Right: Backhead of a small $7\frac{1}{4}$ in. gauge "roundtop" boiler by A. F. Farmer

Below: Fitting the side stays. Note the special long tap wrench

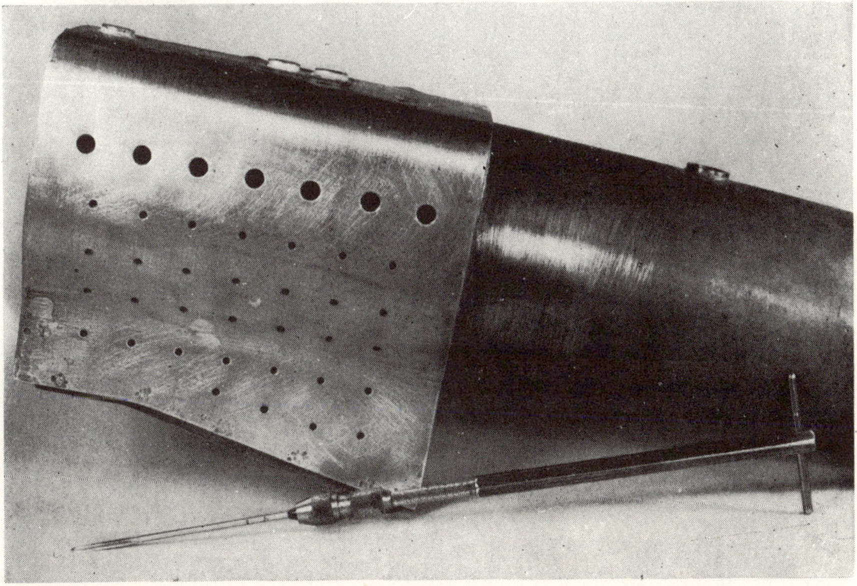

Completed $3\frac{1}{2}$ in. gauge Belpaire boiler by A. F. Farmer

Barrel, inner and outer firebox wrappers of the boiler above

Parts of "Simplex" boiler with wooden formers made by G. H. Corlett of Transvaal

A third method, which can be recommended for mild steel boilers, is shown in Figure 25 (right). A cover strap is put on both the inside and outside of the barrel and the whole riveted up, using four rows of rivets. In this method, the strength of the join depends entirely on the rivets, which must therefore be of a diameter suited to the thickness of the plate and spaced accordingly. Suitable rivet proportions and construction methods for riveted boilers are dealt with towards the end of this chapter.

The brazing of a barrel longitudinal joint in a copper boiler should be carried out as follows: The cover strap is cut to the full length of the barrel less the necessary clearance for the flange of the smokebox tubeplate and for the front flange of the throatplate (if any). It is now annealed by heating to a dull red colour and then plunging into clean cold water. Next the strap is shaped in the bending rolls to the required radius so as to lie snugly inside the barrel. If no bending rolls are available, the strap can be radiused quite easily by gently beating it over a length of round steel bar with a hide hammer.

The strap is placed in position and holes drilled for rivets to hold it in position for brazing. The rivets for this service do not need to be larger in diameter than the thickness of the sheet material; they may be of the usual snaphead type and spaced at about 3 in. pitch. The holes in the barrel are countersunk on the outside.

Assuming that a ready-prepared pickling tank is available, both the barrel and the cover strap are next thoroughly cleaned with coarse emery cloth and then placed in the pickle for about ten minutes, after which they are removed, washed in cold water and dried.

Now the flux for silver soldering should be prepared, according to the manufacturers' instructions; the strap is given a coating of flux, placed in position and riveted, the rivets being driven into the countersinks with the

Fig. 25. Riveting a barrel seam

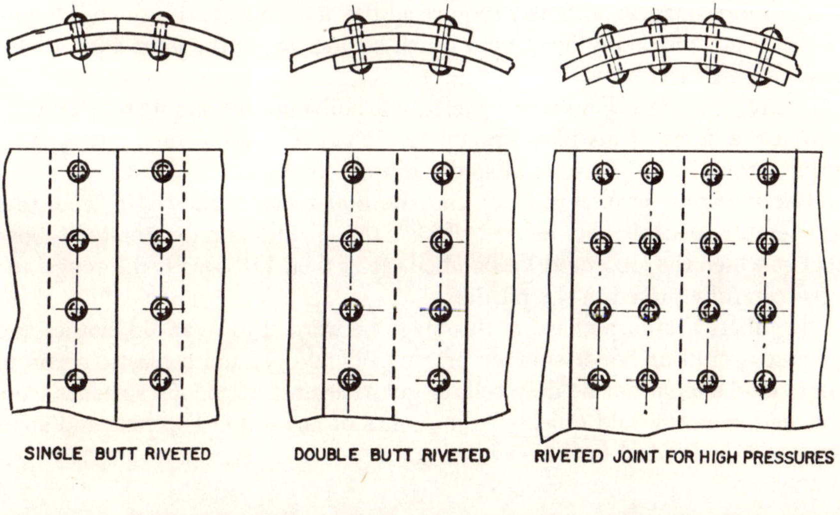

ball end of a medium weight hammer; quite light blows are all that is required in copper work. After a quick rub over with the coarse emery, the joint is then ready for soldering.

Lay the barrel horizontally in the brazing hearth with the longitudinal joint uppermost. Flux along the seam quite liberally, and if using a low-power blowlamp or blowpipe, pile up coke all around the barrel, but not too close to the joint, so that there is no danger of coke particles or dust getting into it.

The flame is now applied, and the aim when using blowlamps, gas or propane blowpipes, should be to heat the whole seam uniformly until the whole of the immediate area is at such a temperature that the brazing alloy in use melts and "flashes" immediately all along the joint.

A stick of the brazing alloy should be held ready in the left hand and this should be dipped into a tin of dry flux before touching it on the barrel joint. The temperature of the barrel must be watched carefully, judging by the colour of the metal. It will first turn black; then as it gets hotter it will gradually turn brighter and very soon begin to glow a dull red colour. At this stage it will be hot enough to melt best grades of silver solder, such as Easyflo, Easyflo No. 2, and Argoflo. If the higher melting point alloys are to be used, such as "Mattibraze", C.4, or B.6., the temperature of the boiler will have to be higher accordingly, the colour being quite a bright red to melt, for instance, B.6. alloy.

The important thing to note is that the boiler joint itself must be at the melting point of the solder before the solder is applied. It is quite useless to apply the solder before this stage is reached as all that happens is that the flame causes the solder to melt off in blobs, forming small balls which tend to run right off the surface of the boiler.

The unskilled operator will however soon discover when he has the seam at the right temperature, as if matters are correct the silver solder will immediately "flash" on contact with the hot barrel, and will run right along the seam, filling every crevice. The higher melting point alloys are however a little more sluggish, and may require a little assistance to run the full length of the seam, this being done by a quick scratch along the joint with a hard steel "scratch rod".

Before passing the joint as completed, the boiler builder should momentarily switch the flame of his blowpipe on to each rivet head in turn, giving each a quick touch with the stick of solder dipped in dry flux.

The blowpipe is now put out and the boiler barrel allowed to cool to a temperature considerably below "black", though still at a temperature above that at which the job could be handled. It is then lifted with the tongs and very carefully placed in the pickle.

Beginners should perhaps at this stage be warned to wear old clothes and a stout overall or boiler suit for brazing operations, and a plastic apron of some kind should not be despised, to guard against accidental splashes from the acid, which would quickly ruin clothes of any kind. Top hats and spats are not exactly ideal for boiler making!

After immersion in the pickle tank for about 20 minutes, the barrel may be lifted out and thoroughly washed and scrubbed with Vim or a similar coarse cleansing powder. It will then be nice and clean and the builder can examine his work for possible flaws or missed places.

Needles to say, if any parts of the seam appear faulty, they must be dealt with properly before proceeding further—by thoroughly cleaning the doubtful spot, re-fluxing, re-heating and re-soldering, in fact the whole sequence of operations must be gone over all over again.

Important—If the oxy-acetylene blowpipe is to be used for this operation, great care must be taken to avoid over-heating and burning the metal, through the very high temperature of the flame. The nozzle of the blowpipe must not be brought too close to the job and the flame must be *kept on the move all the time*, up and down the seam, and as soon as the required temperature is reached, the silver solder is applied and the operation is completed as described previously.

It cannot be over-emphasised that if the oxy-acetylene flame is allowed to remain stationary on copper work, the metal will be over-heated and perhaps even completely melted before the operator realises the danger.

Taper Barrels

Although it is possible to produce tapered barrels from ordinary seamless copper tube, by annealing and stretching the metal, this method is better left to the expert sheet metal worker.

Another method of making taper barrels is to choose a seamless tube of the diameter required at the larger end, cut out a V-shaped piece, almost the entire length of the tube, anneal and then close in the material until the two cut edges meet, after which the seam is dealt with as described previously.

But this is rather an extravagent process, and the author recommends the use of ordinary flat sheet for taper barrels.

It is quite possible to calculate the exact size and shape of the flat sheet to form a taper barrel after rolling, but the process is somewhat complicated by the fact that most taper barrels are arranged with their underside horizontal with all the taper on the top, thus the front and rear ends must be at 90 degrees to the base, rather than to the longitudinal centre-line of the cone.

It is therefore suggested that a dummy taper barrel be first rolled up from thin cardboard, the edges being overlapped and glued together with a quick drying glue. To lay out the cardboard in the flat, the required diameter at the small end and the large end are taken off the drawing, and the circumference at each point calculated by the simple formula: Circumference $= \pi \times D$ or $D \times \frac{22}{7}$. A longitudinal centre-line AB (Fig. 26) is now drawn and the two circumferences CD, EF, set off at 90 deg. at a distance apart equal to the required length of the barrel, plus a small allowance for trimming (about $\frac{1}{16}$in. per inch of barrel diameter).

Lines are then drawn from C to E, and from D to F. Now draw lines EG

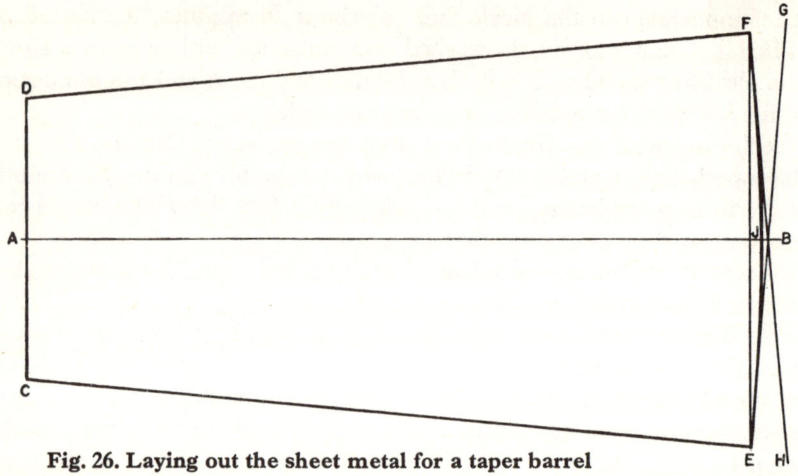

Fig. 26. Laying out the sheet metal for a taper barrel

and FH at 90 degrees to CE and DF respectively as shown. The required edge of material is then given *approximately* by the curved line EJF which can be drawn in by a suitable trammel. There is no need, if working in cardboard, to draw a similar arc at the smaller end.

The cardboard template is now cut out with scissors, allowing an overlap DXYF for glueing; it is then rolled up and glued together. If the taper barrel is one with a horizontal base, the ends can now be squared up with the scissors, checking with a square on a surface plate or other flat surface.

When the cardboard dummy barrel is complete and to correct dimensions, it is cut and opened out and used as a template for marking out the metal sheet. Once again a small cutting allowance may be made on the width of the material, $\frac{1}{16}$ in. being sufficient for barrels up to 4 in. diameter, larger barrels in proportion. Although it is normal to arrange the longitudinal seam of a locomotive barrel made from copper along the bottom, if the barrel is a steel one, the seam should be placed somewhat above the half-way mark.

The ends of a parallel boiler barrel can be squared off in the lathe. Two wooden plugs should be turned up, a close fit in the ends of the barrel. To the centre of one plug, a steel bolt is fitted, and a centre put in this so that the outer end of the barrel can be supported by the tailstock.

The Flanged Plates

Having completed the boiler barrel, the various flanged plates can be put in hand. The flanged plates required for a locomotive boiler are as follows:— the smokebox tubeplate, the firebox tubeplate, the firebox backplate, the throatplate and the backplate or backhead.

The formers for flanging these plates can be made from mild steel, or from hard wood such as oak or beech, or from a combination of the two. A hard wood (teak is also suitable), if the required thickness, will generally stand up to the production of a single flanged plate of each type (the former for the firebox tubeplate generally suffices for the firebox backplate also, according to the design of the boiler). The author generally uses a combination of steel and wood, $\frac{1}{8}$ in. steel being used on the side that is to receive the plate and $\frac{1}{4}$ in. hard wood on the other, the two being held together with glue plus a few countersunk screws.

For a more durable former, $\frac{3}{8}$ in. thick mild steel may be used, but as this would involve much hard work in sawing, an oxy-acetylene cutter would be most helpful. The edges of all plate formers for copper plates should be rounded off at a *minimum* radius of about $\frac{3}{32}$ in., according to the size of the plate; formers for steel boilers however require rather more rounded edges.

The sheet material is cut to size, allowing for the width of the flange and for the thickness of the sheet itself; it is then thoroughly annealed, clamped to the former and the flange beaten over with a hammer. A soft hammer is recommended for this work and the plate must be re-annealed immediately it shows signs of going hard—a thick plate may require annealing three or four times before flanging is complete.

Smokebox tubeplates should always be flanged over formers large enough in diameter to leave the flanges somewhat oversize; the plate may then be chucked in the lathe and turned down to the exact size required to fit closely the inside of the barrel.

In locomotive boiler work, the throatplate of a Belpaire type firebox is always the most difficult plate to make. If the boiler is a small one, $\frac{3}{4}$ in. scale or smaller, and of the all-brazed variety, it is not essential to flange the throatplate forwards to receive the rear end of the barrel, though some additional means can be provided to increase the area of the joint, into which the silver solder can penetrate—i.e. by means of a ring as in Fig. 23. The throatplate must however be properly flanged to the rear, so that the outer firebox wrapper can be secured to it.

For large Belpaire boilers, the throatplate should be double flanged to take both the barrel on one side and the wrapper on the other. A metal former is first cut out on its outside only for the flanging on the wrapper side of the plate, after which the partly completed throatplate is removed from the former and a hole cut in the former in the required position and to a diameter slightly greater that the *inside* diameter of the barrel at the firebox end. This can be done in the lathe, the former being held in the 4-jaw chuck.

A circular hole is now cut in the throatplate of a diameter equal to the hole in the former less twice the flanging allowance. After further annealing, the partly flanged throatplate is again clamped to the former and flanged forwards, into the hole cut in the former. Finally the throatplate is returned to the lathe for light turning to receive the rear end of the barrel.

Throatplates for round-top boilers can be made in much the same way, though as mentioned previously, if the firebox wrapper is made from an extension of the barrel, only a very simple throatplate is required—provided that the joint between the throatplate and the barrel can be brazed.

The two firebox flanged plates can often be flanged over the same former. The firebox tubeplate and firebox backplate are often of the same shape and dimensions, though the tubeplate is sometimes deeper, so that the former for flanging these plates should be made deep enough to allow for this. Many G.W.R. and L.M.S. type fireboxes, however, narrow in towards the rear, necessitating separate formers.

After completing the flanged plates, the boiler maker generally tackles the outer wrapper of the firebox. The wrapper is cut to length and to estimate the required width of material needed, a good dodge is to obtain some soft iron wire of the same thickness as the sheet metal to be used, bend this around the firebox tubeplate, cut it to length, then straighten it out and measure its length. A small additional allowance may be added with advantage—about $\frac{1}{16}$ in. per 6 in. should be sufficient.

The outer wrapper is now bent to shape, using the bending rolls and checking against the flanged plates at frequent intervals. If no bending rolls are available, round bars of mild steel of suitable diameters can be used, clamped in the bench vice and supported at the outer end as required. The material should of course be well annealed before any bending is attempted.

The wrapper is next fitted around the throatplate and backhead, though the backhead is not of course attached permanently at this stage. Snaphead rivets are generally used to hold the wrapper closely to the rear flanges of the throatplate. It should not be thought that a large number of rivets around the wrapper is a desirable thing on the grounds of strength; on the contrary, the strength of the joints of the firebox is derived from the silver-soldering process rather than from the rivets. Every hole drilled for such rivets is after all a potential source of leakage.

At this stage, the joint between the outer wrapper and throatplate and between throatplate and barrel can be brazed. Brazing alloys such as B.6. and C.4 can be recommended here, though apart from the greater expense, silver solders such as Argoflo or Easyflo No. 2 give an even better result, due to their better penetrating properties. A good fillet of the solder should be run all around the joints on both sides, special attention being given to the underside of the barrel, up against the throatplate. If the penetration is sound, it should be possible to see a silver line all around the inside of the firebox. After the usual "pickling" and a thorough wash, attention can now be given to the inner firebox.

The firebox tubeplate is marked out for the tubes and flues, deeply centre-popped and the holes pilot drilled. This plate can now be used as a drilling jig for the holes required in the smokebox tubeplate. When clamping the smokebox tubeplate to the firebox tubeplate for this purpose, any rise in the level of the tubes towards the smokebox end of the boiler must be allowed for.

The tube holes are next opened out to a shade under their finished size. It is a good plan to partly finish these holes with reamers; the reamer should however on no account be put right through, but only about half-way along the "lead", so that the hole is still left a few "thou" undersize.

As the diameter of the holes required to take the superheater flues may well be beyond the capacity of the builder's drilling machine, recourse is often made to the lathe. It is generally possible to clamp the tubeplate to the lathe faceplate (using wood packing to keep the plate a short distance clear of the faceplate to allow for tool "run-out"), and to bore the holes using a normal boring tool clamped under the tool-holder. Once again, the holes should be left a few "thou" undersize.

The firehole ring may now be made up and fitted to the firebox backplate. A good method of making the firehole ring is to turn it from thick-walled copper tube, both ends of a short length of this being turned so as to form a step in the middle. This can then be put through a hole of suitable size cut in the firebox backplate, flanged over on the inside and brazed.

If an oval firehole is preferred, the best approach is to first turn the ring in the lathe as described previously, anneal and bend to an oval shape in the vice, then mark off and cut a hole in the firebox backplate to suit. The ring is then fitted, riveted over and brazed as before.

If the tubes and flues are to be silver soldered to the firebox tubeplate (and this is strongly recommended in a copper boiler), the holes in this plate may be lightly countersunk on the outside—that is the side nearest to the smokebox—which assists the formation of a sound fillet of solder..

The firebox inner wrapper is now cut and shaped to match the flanged plates as described for the outer wrapper, and sufficient snaphead rivets put in to hold the wrapper closely to the flanges all around the firebox. In some narrow-type fireboxes, it may be found difficult to get a riveting "dolly" inside in certain positions where rivets are needed. The solution here is to make up gunmetal or copper screws, and drill and tap the wrapper and flange, which must be kept in close contact while the screws are being put in. As mentioned previously, the number of rivets and screws used should be kept to the minimum, subject to the wrapper being held closely to the flanged plates.

At this stage, a decision should be taken as to whether the firebox stays are to be silver soldered or caulked with a high melting point soft solder. *This decision makes a great deal of difference to the method of assembly of the remainder of the boiler.*

To take the all-silver soldered assembly first, the firebox backplate must not at this stage be silver soldered in place. It may, however, be inserted and held in place temporarily by screws to keep the firebox as a whole in shape.

Crownstays

If plate crownstays are being adopted, these can now be made up and riveted to the top of the firebox, once again using the minimum number of snaphead rivets, when the whole unit may be brazed using an alloy such as B.6, C.4, "Mattibraze" or "Argobond". The exact height of the plate crownstays may be checked by temporarily assembling the inner firebox inside the outer wrapper and any adjustments required made before carrying out the brazing operation. Care should be taken that the brazing material flows throughout all the joints, especially along the base of the crownstays, over all screw and rivet heads and around the firehole ring if this has not been dealt with previously. The flame should be kept away from the area of the tube holes as far as possible or this part of the firebox may become overheated. As with all brazing and silver soldering operations, the job must now be pickled in the acid bath, washed thoroughly and carefully examined before the next part of the boiler is attempted.

If a fusible plug is to be fitted, this should be done at this stage, the bush for the plug being silver soldered into the crown near the front of the firebox and clear of any stays.

Fitting Tubes to Firebox

It is now time to fit the tubes and the flues and as mentioned previously, it is advisable to silver solder these to the firebox tubeplate. At one time, it was customary in the smaller locomotive boilers to thread the ends of the tubes and screw them into the tubeplate, afterwards caulking with soft solder. This method is a poor one; for one thing, the tubes which must be of fairly thin material if they are to be efficient, are dangerously weakened by threading, while the tubeplate is not thick enough to take more than a few threads unless specially thickened locally.

Some builders screw their tubes into the tubeplate and then silver solder them. This is better but still does not entirely overcome the objections mentioned and in any case, this method involves a great deal more work.

The smokebox tubeplate can be used as a spacer to hold the outer ends of the tubes while they are being silver soldered into the firebox tubeplate. Some boilermakers complain that while in the middle of the silver soldering operation, one or two of the tubes slip right through the tubeplate thus spoiling the whole job. This can be avoided if the tube holes in the tubeplate are left undersize as mentioned previously and the tubes put into the lathe and given a very light skim for a length slightly greater than the thickness of the tubeplate, leaving the ends a tight fit in the tubeplate, with a shoulder on the tube which will effectively prevent them slipping through. The tubes and flues should, of course, be annealed and thoroughly cleaned before fitting and *it is an excellent plan to file a few shallow nicks (not deeper than* 0·010 *in.) in the tubeplate holes to assist the silver solder to penetrate.*

In this first method of boiler assembly, the firebox backplate must not be silver soldered to the firebox wrapper until later on, as will be explained shortly. Probably the easiest way to carry out the silver soldering of the tubes to the tubeplate is to stand the whole assembly up on end with the tubes pointing straight up in the air. Plenty of flux should be used and the silver solder (Easyflo or Easyflo No. 2 are excellent for this purpose) preferably in wire form, wrapped around the base of each of the tubes hard against the tubeplate.

The flame from the blowpipe can be applied from inside the firebox, though it must be kept continually on the move to prevent local overheating. On bringing the whole to a dull red heat, so that the solder runs freely, more solder can be fed in from all around before the flame is removed. After pickling and washing, the assembly should be carefully examined. If the silver soldering has been properly done, it should be possible to see a silvery ring all round each tube *inside* the firebox, showing that penetration is complete. If any tube ends look doubtful, the offending plates should be carefully cleaned, re-fluxed and heated again, working from the inside, the front ends of the tubes being held firmly by soft iron wire in case any of the tubes should slip out during the re-heating process.

The inner firebox/tube assembly can now be tried inside the barrel/outer wrapper assembly and any adjustments necessary made to the crownstays (if of the plate girder type).

The Foundation Ring

The foundation ring is normally made of four separate sections of bar material and a ring built up in this way is certainly an easier assembly problem than a one piece foundation ring—generally used in full-size practice. A length of bar should now therefore be cut to form the front section and its lower edges bevelled to allow better access for the brazing alloy. It is then temporarily clamped into position between the throatplate and the firebox tubeplate and at the same time the flanges of the crownstays can also be temporarily clamped to the underside of the outer firebox wrapper. If the crownstays are to be of the rod type, a careful check should be made that the distance between the top of the crown and the outer wrapper is correct according to the drawings.

A careful examination of the whole boiler should now be made, special attention being paid to the correct clearance between the bottom row of tubes and the bottom of the barrel and the water spaces at each side of the firebox (as viewed from the rear of the boiler). If everything is in order, the front section of the foundation ring is riveted up using as few rivets as possible (two is usually sufficient in small boilers), after which the two side sections of the foundation ring are cut and fitted and riveted into position.

The flanges of the crown stays are now riveted to the outer wrapper. Some kind of special tool is useful to get the rivets into position from the inside, as those towards the front of the firebox will be too far in to be reached by the hands except in very large model boilers. For small boilers, a long strip of copper about $\frac{3}{8}$ in. by $\frac{1}{16}$ in. section with a vee slot cut in the end will be found to hold the rivets sufficiently firmly to get them through their holes if the holes are nicely to size. The rivets should again be of the snaphead variety.

To support the heads of these rivets adequately, especially those in the two outer rows, requires careful thought and some simple home-made appliances (see Figs. 18 and 19).

If it is not intended to silver solder the flanges of the crownstays to the outer wrapper (though this is strongly recommended in smaller boilers) sufficient rivets must be put in to ensure adequate strenth.

The smokebox tubeplate can now be given a final clean up and the tube and the flue holes lightly countersunk on the outside. It is then pressed into position in the barrel with the tubes standing through the plate for an amount about equal to one-quarter of the tube diameter (the length of the flues should not be greater than the tubes) and the flange of the tubeplate should be on the inside so that a convenient groove is left all around the plate—helpful when silver soldering. A few copper or gunmetal screws are put in through the sides of the barrel to hold the tubeplate firmly while the tubes are expanded, this being done by a couple of taper drifts of suitable diameters. The drifts should be polished and lightly greased before use to facilitate withdrawal.

The smokebox tubeplate can now be silver soldered using either best-grade or second-grade silver solder as preferred, but before doing so it is not a bad scheme to flux the crownstay flanges ready for soldering these immediately after the smokebox tubeplate has been dealt with.

To silver solder the smokebox tubeplate, the boiler should be arranged on end with the smokebox end at the top. Users of blowlamps and burners of limited capacity will find it easier if the boiler barrel is put through a hole cut in a sheet of steel or in the bottom of the brazing hearth; coke can then be piled up all around the barrel to within half an inch or so of the top to conserve the heat. As soon as the silver soldering of the tubeplate and tubes is completed, the boiler can be shifted onto its back with the rear edge of the outer wrapper overhanging the brazing pan and the flame of the blowpipe transferred to the tops of the crownstays.

It will be found easier to silver solder the crownstays to the wrapper if a thin strip of the solder, well coated in flux, is laid against each of the flanges of the crownstays for their full length; the flame can then be applied from underneath and the solder watched carefully until it "flashes" and runs freely right along the joint.

Now is the time to tackle the fitting of the stays and these will consist of the following varieties: the short stays between the throatplate and the firebox tubeplate, the stays between the inner and outer wrapper, and if girder type crownstays are not being used, the rod stays between the crown of the firebox and the outer wrapper.

The position of each stay should be carefully marked out and after centre-popping and drilling to tapping size, the drilling should be continued right through into the inner wrapper, taking care that the drill is kept as near as possible vertical to the inner wrapper, otherwise it will wander and possibly break. As far as the rod-type crownstays are concerned, the holes in the outer wrapper are then opened out to the clearing size for the stays to be used. The tapping of the holes in the inner wrapper is best done from the outside so as to preserve correct alignment; in any case, many of the outer rows could not be reached from the inside. In large boilers, it may be found that a special long tap is needed to reach the inner wrapper, but if such a tap is not available, it is not difficult to extend a standard tap by brazing on a suitable extension.

The rod stays, which may be made of copper, gunmetal or monel metal, are machined in the lathe so that only the ends are threaded, the plain portion between the threads being reduced until a few "thou" below the core diameter of the thread. This practice, contrary to popular belief, does not reduce the strength of the stay when in tension. The stays are screwed into the inner wrapper through the clearing holes in the outer wrapper, then locknutted at both ends and finally run over with best-grade silver solder.

Belpaire fireboxes require transverse stays to support their flat sides and these can also be put in at this stage. When drilling for the side stays, care must be taken that the boiler is held at the right angle under the drilling machine so that the drill is fed square into the plate. In certain positions on the sides of a locomotive firebox, it is not possible to get the drill exactly at 90 degrees to both the outer and inner wrapper at the same time, so that a compromise must be adopted.

It will be appreciated that while drilling and tapping for all these stays, the firebox backplate must be kept in its correct position (and this can be done by means of a couple of small toolmakers' clamps) otherwise the inner wrapper may be badly distorted and the backplate will not fit.

For those stays between the throatplate and the firebox tubeplate, an extra long tap holder will be required, as a normal tap holder would foul the barrel as it was being turned. This special tap holder is quite easily made; the author uses an Eclipse chuck-type pin vice which is pressed into a long length of hexagon bar with a tommy bar put through the outer end at right angles.

If gunmetal or monel metal is used for any of these stays, the boiler builder should not try to rivet the head over on the outside. Riveting the heads is sound practice for copper stays, though even these should be annealed if nice heads are to be obtained without damaging the sheet metal. Recommended types of stays are shown in Fig. 7. For the ordinary firebox side stays, after tapping right through both plates, the outer hole is opened out and lightly countersunk on the outside. The stay is turned in the lathe, the thread being machined in the usual way and the material is then *partly* parted off so that we have the rest of the bar material with which to screw the stay right home, after which the stock is twisted off. After silver soldering, the head can be rounded off a little with an old file and coarse emery cloth.

When the builder is satisfied with the staying of the sides of the firebox, the throatplate and the rod type crown stays, if fitted, the firebox backplate can be silver soldered in place. Care must be taken that the firebox backplate does not shift while the heat is being applied; although it is next to impossible to rivet or screw it in the inner wrapper (as the outer wrapper is in the way) it can be held firmly enough by a couple of small clamps made of rusty steel strip, or if the builder has two small toolmakers' clamps that have seen better days and are not quite up to general work any longer, they would do fine. The reason for saying that these clamps should be rusty is that if they were not, the silver solder might run down and adhere to them and it might prove a very difficult job getting them off again.

The plate should now be pickled, washed thoroughly and fluxed before inserting in place and it is not a bad plan to smear the lower parts of the flange with "Easyflo paste" so that even with the clamps in the way, the silver solder should "take" over the whole area of the flange. But no chances should be taken over this. The flame should be kept away from the clamps as far as possible and when the whole backplate starts to glow red, the flame should be moved round, starting from the top and going round first the left-hand side and then the right-hand side.

The upper part of the backplate should now be allowed to cool just enough to "set" the solder. The two clamps are then whipped off, using pliers on the tightening screws, and the flame is played on the bottom corners of the plate where the clamps were. The corners should be touched with a stick of "Easyflo" dipped in the dry flux and the builder should then have no difficulty in getting a sound joint, especially as the plate is only a little below the level of the edges of the outer wrapper, so is quite accessible.

At this stage, the backhead can be completed. All bushes are fitted, pushed in from the inside with the exception of that for the regulator which is best put in from the outside, and they are then brazed. The hole is cut for the firehole ring, the position and shape required being checked off from the ring itself. The remaining, rear, section of the foundation ring is also fitted into place and secured with a couple of rivets. A few gunmetal screws may be found necessary to ensure that the backhead and the outer wrapper are closely in contact all the way round, but no more than are absolutely necessary should be used and they should be given a touch with the silver solder before completing the joint, just in case.

As the firehole ring has to be flanged over on the outside, care should be taken when doing this that the position of the backhead is not altered in relation to the outer wrapper, nor the backhead itself distorted. Proper support should, therefore, be given on the inside of the firehole ring.

There now only remain the stays between the backhead and the firebox backplate to be fitted. Builders who have no oxy-acetylene equipment may find it very difficult to silver solder these stays but if they are carefully threaded in both plates, it is quite safe to use high melting point soft solder. The last major brazing operation—the foundation ring, backhead flange and firehole ring—generally requires the greatest amount of heat, so unless oxy-acetylene equipment is available, care will have to be taken in packing around coke, etc., to conserve the heat as much as possible. First, asbestos sheet should be put inside the firebox to protect the crown and the tubes. Short pieces of asbestos sheet should be softened with water and rolled up so as to fit inside the tubes at the firebox end; another piece of asbestos sheet is then laid against the tubes and a further piece across the inside of the crown of the firebox. On top of this, coke can be built up both inside and outside the firebox up to about one inch from the foundation ring—the boiler being laid on its back at this stage.

Should difficulty still be encountered in reaching the required temperature, there is no objection (other than expense) in using best-grade silver solder around the foundation ring, though it will be found that "Easyflo No. 2" or "Argoswift" or "Argobond" will provide a better fillet and be less likely to run right through any small cracks or holes left between the various sections of the foundation ring. The flame of the blowpipe should be worked right round the foundation ring to ensure a good joint and while the boiler is still hot, it can be set up on end with the backhead on the top so as to deal immediately with the backhead flange and the firehole ring.

Any bushes required other than those in the smokebox tubeplate and the backhead are next pressed into position. These will include bushes for safety valves, top feed fittings, check valve bushes (where these are situated on the sides of the barrel) and bushes for manifold fittings on the top of the firebox. These bushes should be silver soldered using best-grade solder and it should be possible by suitable screening off with firebricks, etc., to fix all these bushes without any danger of melting some other joint in the boiler.

Boilers with Caulked Stays

We can now retrace our steps to the type of boiler assembly where the firebox stays are to be threaded and caulked with "high melting point" soft solder rather than silver solder.

In this type of assembly, it is the usual practice to complete the silver soldering of the whole of the inner firebox, and having fitted this to the barrel/outer wrapper assembly (before any stays are fitted) the foundation ring is completed and the backhead fitted in position and silver soldered. Thus, we have a complete boiler assembly with no holes at all at this stage. If the builder now fits and silver solders the various small tapped bushes in place, it will be possible to give the boiler a preliminary test. This test must not, of course, be done to a higher pressure than about 20 p.s.i. as otherwise the flat plates might be distorted.

If this preliminary test shows up any leaks, these should be dealt with immediately by silver soldering, and as soon as the boiler has been found to be quite pressure-tight, a start can be made on fitting the remaining bushes such as a large dome bush or a large regulator bush and then finally the stays.

In this case, all these stays must be carefully threaded through both plates using a long stay tap if available. The stays are then screwed in tightly and are secured on the inside of the firebox by nuts which should for preference be somewhat thicker than the length of stay protruding, so that the ends of the stay are protected to some extent from the direct heat of the fire.

The reader will now appreciate that in the first method of assembly, the reason for leaving the fitting of the firebox backplate and the backhead while the great majority of the firebox stays were fitted, was done so as to permit the use of the blowpipe inside the firebox without the flame of this being put out by lack of air.

Returning now to the next stage in boiler construction, when all staying has been completed and all bushes properly fitted, the builder should be ready for the hydraulic test.

The Hydraulic Test

The hydraulic test on any type of boiler is carried out as follows: All the bushes are blanked off, including the regulator flange and the main steam pipe flange on the smokebox tubeplate. A reliable pressure gauge which must read to two and one-half to three times the working pressure of the boiler is attached to a convenient bush in the upper part of the boiler, i.e. a safety valve or similar bush. A check valve is screwed into one of the lower bushes, such as those provided for blowdown valves or one of the bushes provided for the feed water. The boiler is now completely filled with cold water, all air being excluded, and it is then ready for the first test.

The necessary pressure for the hydraulic test can be supplied by any suitable hand pump; the emergency hand pump often fitted to the tender or a side tank of a model locomotive is quite satisfactory for the purpose. A few strokes of the pump will bring the pressure up very quickly and if this is a boiler in which the firebox stays have been caulked by soft solder and to which an initial test to 20 p.s.i. has been applied, the pressure can be raised gradually to twice the designed working pressure of the boiler. It should be held there at least 10 minutes while the boiler is carefully examined for deformation or leaks. In the case of large boilers (above 1 in. scale) it is advisable to maintain the test pressure for at least 20 minutes. It is also an excellent idea to apply the pressure and then allow the pressure to be released a number of times in succession; this will soon disclose if there are any fundamental weaknesses in the boiler.

Deformation of the boiler may be a serious matter, and should be attended to. In a steel boiler, this could be due to a fault in design, to bad workmanship, or to both. In copper boilers, some slight bulging may occur which may not be serious. After brazing operations, the copper will be very soft, and it will

Fig. 27. The hydraulic test assembly

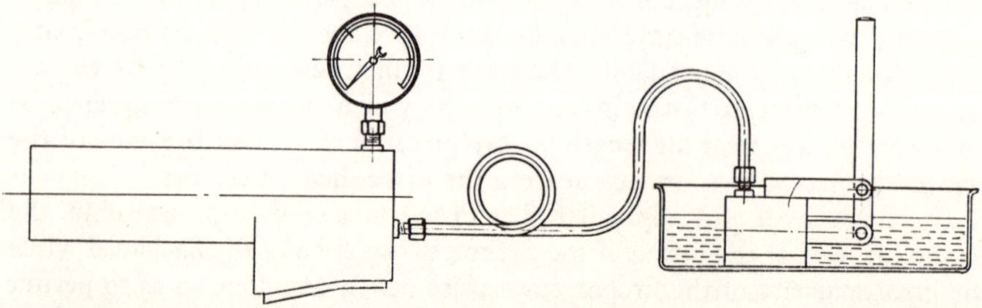

bulge rather easily if given the opportunity. If the bulge is small, it may be gently tapped back by the use of a hide hammer, but great care must be taken to avoid straining one of the stays, which may be in the locality. This careful hammering will "work-harden" the copper, and the boiler will then stand a considerably higher pressure without trouble.

The boiler builder should not however be upset if he finds several small leaks on the first hydraulic test of a new boiler. Even quite experienced workers often find a small leak or two: many leaks are easily cured.

An important difference should be noticed here: small leaks in copper boilers rarely cure themselves, so they should be attended to immediately. Small leaks in steel boilers however very often cure themselves after a first test under steam, or perhaps after two or three steamings. But obviously, even in a steel boiler, if the leak is so bad that difficulty is experienced in raising an appreciable pressure in the boiler, the leak must be properly dealt with.

It is not possible to say here what is the best method of curing a leak in a model boiler: it depends a great deal on the position and size of the leak. A small pinhole (which could be ignored in a steel boiler) can be quickly cured in a copper boiler—provided that it is accessible—by drilling and tapping the metal, say 10 BA, using a taper tap, and screwing in tightly a short stub of copper threaded to match, the threads being smeared with ordinary white plumbers' jointing before screwing in. A bad leak between the end of a tube and its tubeplate is generally better cured by thoroughly cleaning the spot, including a short spell in the pickle tank and thorough washing; the metal should then be fluxed and heated with a fine but hot flame until a smallbead of best-grade silver solder can be run in, which would make a sound and lasting cure.

One of the most difficut leaks to cure is in a combustion chamber, where it may prove impossible to reach, to use either of the methods described previously. A solution which has been used with success—though it is one which is only suggested as a last resort—is to heat up the whole boiler to the melting point of soft solder (the higher melting point soft solder should be used); pour a few spoonfuls of the appropriate flux inside the boiler and then put inside as near to the offending place as possible a few small sticks of the soft solder. The boiler is then re-heated a little to ensure that the solder has melted properly. It is then picked up using asbestos gloves and shaken about to try to get the molten solder to fill any tiny holes that may exist. Excess solder can be emptied out afterwards and the boiler thoroughly washed out to remove all traces of the flux.

To cure a bad leak in a steel boiler, there is nothing to beat the oxy-acetylene blowpipe which can be used to run in welding alloy wherever required.

Steam Test

After a satisfactory hydraulic test, the boiler should be given a steam test. Some of the water should be emptied out of the boiler until the water level is approximately at the normal working height and some form of heat applied in the firebox. It is not essential at this stage to start a proper coal fire in the firebox. A gas blowpipe or paraffin or bottle gas burner may be used. Pressure should be raised gradually until about 20 per cent higher than the working pressure of the boiler and it should be held for at least 15 minutes. A further test to the same pressure can also be made after all the fittings have been attached to the boiler.

Before leaving the subject of boiler testing, it cannot be too strongly emphasised that a reliable pressure gauge must be used in such tests; the miniature steam pressure gauges normally fitted to model locomotives are quite unsuitable for testing purposes. Regular hydraulic tests must also be carried out on locomotives in service. Proper tests are generally arranged by the larger model engineering societies and certificates issued; in fact, most clubs insist on locomotives having proper boiler test certificates before they are allowed to run on the club track or anywhere in public.

Other Methods of Construction

So far we have only considered the building of copper boilers having all joints silver soldered. It is, however, often asked whether it is possible to build a boiler in copper without silver soldering or brazing being used at all. The answer is that this *can* be done, but it calls for much better workmanship and careful siting of rivets as well as neat and accurate riveting, as such a boiler depends for its strength entirely on the rivets used in its construction. But it must be remembered that even with perfect riveting, some caulking will be necessary and for this the high melting point soft solder should be used.

The great disadvantage of the riveted and soft soldered boiler is the possibility of permanent damage should the water level be allowed by accident to fall below the level of the crown of the firebox. A compromise is therefore worth considering: silver solder all the joints of the inner firebox and also silver solder the tubes into the firebox tubeplate and rivet the remainder, relying on high melting point soft solder to ensure steam-tightness.

Riveted Joints in Boilers

As mentioned previously, a solid-drawn tube makes much the strongest boiler barrel, while if a barrel is made from sheet metal with the longitudinal joint made in the form of a simple single-riveted lap joint, the strength of such a barrel should not be taken as equal to more than half that of the seamless barrel.

While a single-riveted lap joint is acceptable for low pressure model boilers, for boilers working at pressure from about 50–100 p.s.i. a butt joint with a single strap and a double row of rivets on each side of the gap is much to be preferred. For the highest pressure, the butt joint with double strap is recommended. Fig. 28 gives full details of suitable rivet diameters and spacings.

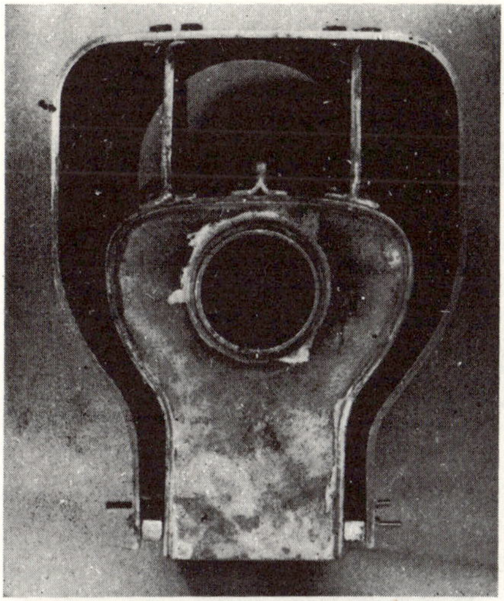

Top left: "Firefly" boiler under construction

Top right: An all-sifbronzed firebox

Left: A simple throatplate drilled for stays

Bottom left: Firebox for "Boxhill"

Below: The "Firefly" firebox backplate

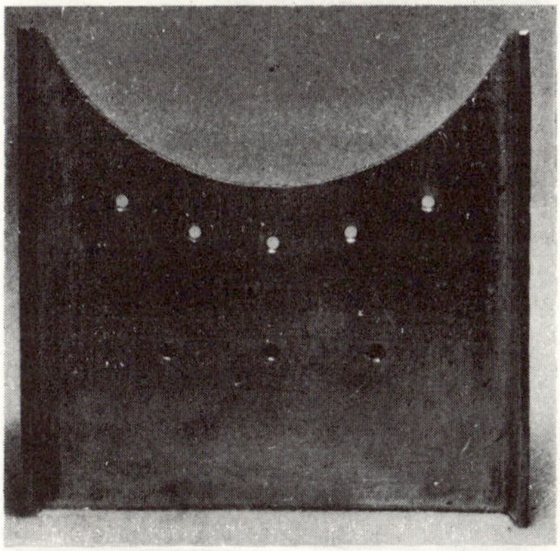

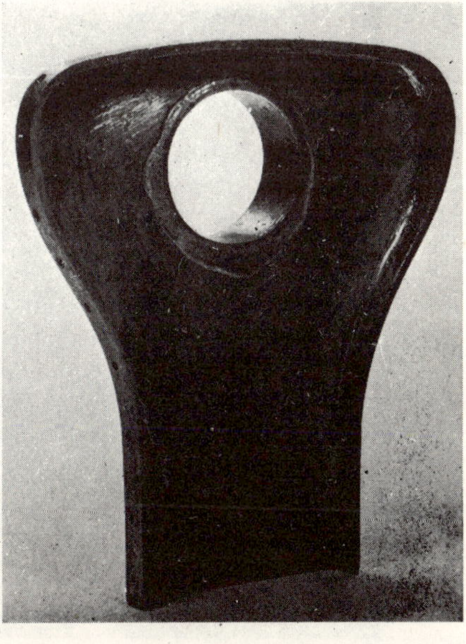

Another view of a properly flanged throat-plate in copper. This was made by Mr. A. F. Farmer for the author's $7\frac{1}{4}$ in. gauge "Highlander"

The smokebox tubeplate on a large model—Mr. J. N. Liversage's $10\frac{1}{4}$ in. gauge "Lord Nelson"

A model locomotive boiler with all joints sifbronzed

Left $3\frac{1}{2}$ in. gauge boiler with barrel joint "joggled" by G. M. Cashmore

Below: brazing the crown stays to the outer wrapper. A picture taken in the old M.E. workshop

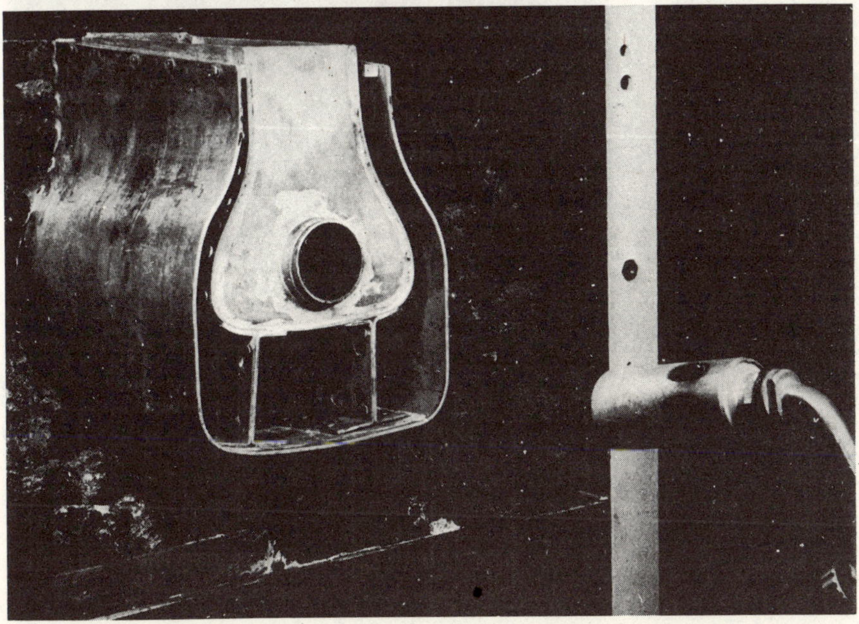

"Too many cooks . . . ?" E. C. Dearman in Australia with helpers

How to turn a smokebox

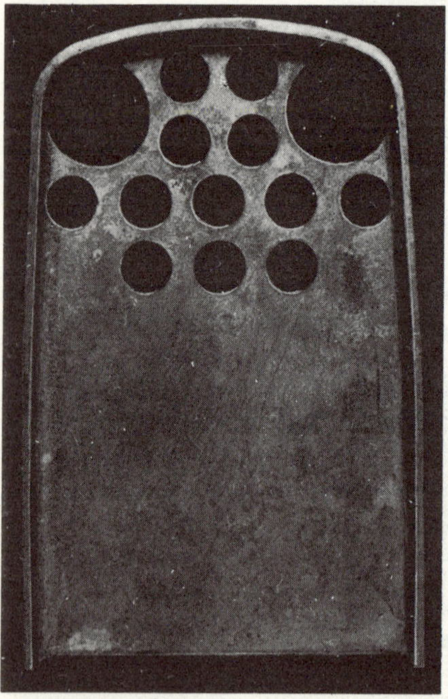

The firebox tubeplate for "Boxhill"—5 in. gauge "Terrier" locomotive

Right: Large Sievert propane blowpipe in use on a wide firebox boiler: Photograph courtesy W. A. Meyer Ltd

T	D	P	L
1/16"	1/8"	3/8"	3/16"
3/32"	5/32"	1/2"	1/4"–7/32"
1/8"	3/16"	9/16"–5/8"	5/16"
1/4"	1/2"	1 1/4"	3/4"
5/16"	5/8"	1 1/2"	15/16"
3/8"	3/4"	1 3/4"	1 1/8"
7/16"	13/16"	1 7/8"	1 7/32"
1/2"	7/8"	2"	1 5/16"

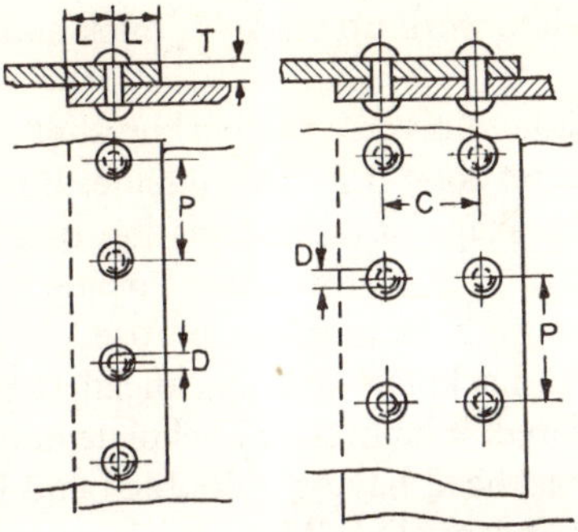

T	D	P	C
1/4"	1/2"	2"	1 1/8"
5/16"	5/8"	2 1/2"	1 1/4"
3/8"	3/4"	2 5/8"	1 5/16"
7/16"	13/16"	2 3/4"	1 3/8"
1/2"	7/8"	2 7/8"	1 1/2"

Fig. 28. Rivet proportions for boiler work

Although countersunk rivets are permissible in barrel seams, generally speaking snaphead rivets only should be used in boiler work; copper rivets in copper boilers and charcoal iron rivets in steel boilers. Holes for rivets should not be drilled too closely to size and the holes should be very lightly countersunk on both sides before commencing work.

As stretching takes place along a boiler seam during riveting, it is a good plan to use close fitting steel bolts put into alternate holes along the seam to hold the edges in correct alignment while the rivets are being closed. Ordinary screws must not be used for this purpose as they do not entirely prevent stretching and may prove difficult to remove. After rivets have been put into position and closed in all the holes in between the bolts, the bolts are removed one by one and rivets inserted.

An important point to note in boiler riveting is that the rivets must be properly supported. The use of a heavy overhanging bar held in the bench vice can only be recommended for very small boilers where copper rivets up to $\frac{1}{16}$ in. diameter will be employed. But in bigger boilers, a proper rivet jammer should be used; one fitted with left and right-hand threads is most convenient. Precautions should be taken, particularly in copper boiler work, that the end of the rivet jammer, opposite to the rivet being dealt with, does not damage the side of the barrel or firebox as the case may be, against which it is pressing.

In building riveted boilers, it may be appreciated that many of the methods used in full-size boiler work cannot be used in models. When fixing a backhead for instance, it is impossible to obtain access to the inside of the boiler to close the rivets if the backhead is arranged in the usual way with the flange on the inside. It is, however, possible to secure it by screwed rivets, though this is certainly not recommended for copper boilers unless the joints are to be silver soldered afterwards.

Screwed rivets are also useful in the corners of the foundation ring in steel boilers in positions where it is not possible to use an ordinary rivet. They can be turned from bar and threaded in the lathe. The head is partially formed and left with a small neck so that when screwed right home in the boiler, a little extra force will shear the bar from the rivet. Screwed rivets should be threaded 32 t.p.i. for plate work up to $\frac{3}{16}$ in. thick and 26 t.p.i. for plates from $\frac{1}{4}$ in. to $\frac{3}{8}$ in. thick.

It is, of course possible to reverse the backhead of a steel locomotive boiler so that the riveting can be done from the outside. But this makes for a rather ungainly backhead. The shape of the foundation ring becomes very awkward and it is more difficult to arrange the fittings neatly on the backhead.

Although the turned firehole ring can be used in a riveted boiler, if the flanging for the firebox backplate and backhead is properly done, another method may be considered. The firebox backplate and the backhead are both dished outwards, the backhead having a double bend in it; the two plates are then brought together and riveted. The final joint is thus only equal to the thickness of the two plates. The usual firehole ring is dispensed with and the rivets are well out of the way of the heat of the fire (see Fig. 9).

The smokebox tubeplate of the riveted boiler is normally arranged with its flange outwards—the opposite to the brazed boiler—so that the rivets are readily accessible.

Steel Boiler Construction

As mentioned earlier, steel boilers are generally used for locomotives of 2 in. scale and larger. A good quality acid open hearth steel sheet should be chosen for boiler construction. When possible, a steel with the following approximate composition should be used; Carbon 0·5 to 0·25 per cent, Phosphorus not more than 0·04 per cent, Manganese 0·40 to 0·45 per cent, Silicon 0·02 to 0·03 per cent, Sulphur not more than 0·04 per cent, balance iron.

Boilers made from mild steel can be made in a similar way to those built from copper (with the proviso that the radii of the flanges of the various plates are made larger, as mentioned previously—this is very important), or they can be entirely riveted. They can also be riveted just sufficiently to hold all the various parts together, the joints being then welded. The last method is

Fig. 29. Recommended flanging radii for boiler plates

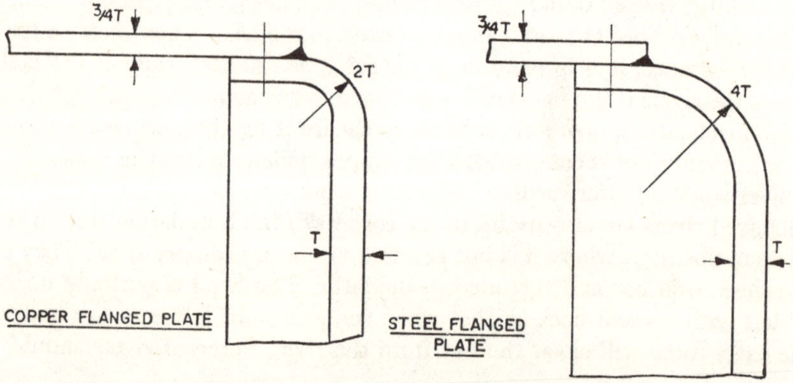

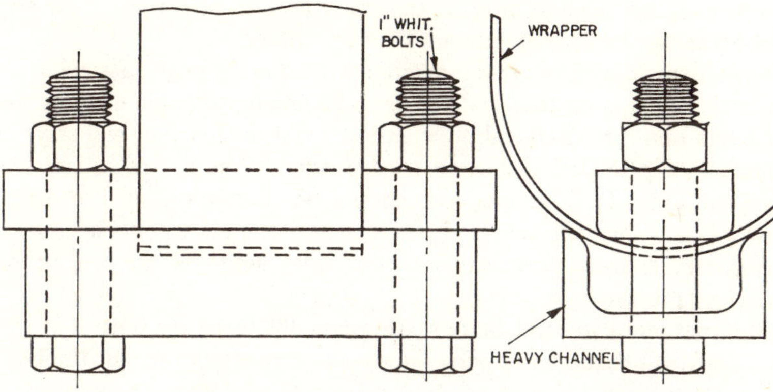

Fig. 30. How to bend a heavy firebox wrapper

probably the best for the amateur boiler maker, though for this oxy-acetylene or electric arc welding equipment is essential.

If flanged steel boilers are welded, they must be butt jointed, not lap jointed. In edge welding a lap joint there is no penetration of the lap as in brazing. The weld is subject to bending fatigue stress and the lap traps corrosion products between the plates, where it cannot be examined. When boilers are to be welded, they must be designed for it, rather than using welding as an alternative to brazing.

Difficulty is sometimes experienced in bending the various wrappers when steel is used for model boiler construction. However, even quite thick material may be dealt with if the method shown in Fig. 30 is adopted. Some really heavy engineering steel channel section should be obtained, drilled for 1 in. bolts as shown, and a heavy rectangular bar drilled to match. If the corners of the bar are well rounded off, and all edges greased before use, no difficulty should be encountered.

In all kinds of steel boilers, proper flanged plates are recommended, and where a fully riveted construction is adopted, the rivets should be arranged as in Fig. 28.

Opinions differ as to the best way of fitting the tubes to the firebox tube-plate in steel boilers. Many builders favour the traditional method of expanding, but it is sometimes claimed that this leads to distortion of the tubeplate. A good alternative is to expand the tubes just sufficiently to ensure that they are quite tight in the tubeplate and then weld them. The welding should be carried out from the bottom of the tube upwards, first to one side, and then to the other, lapping over the end of the first weld.

At the smokebox end, expanding generally gives a good joint.

Some boiler makers favour the use of a copper inner firebox in a steel boiler on the grounds of better conductivity and an absence of trouble through corrosion. If a copper firebox is used, this should be entirely silver soldered; copper tubes should be used, silver soldered into both plates, while the foundation ring and firehole ring (if any) may also be of copper. Side stays in such boilers can be of monel metal with advantage.

Some problems may arise in connection with staying in an all-steel boiler. Few authorities agree on this matter, some favouring wrought iron or steel stays, others a non-ferrous metal. The author's view is that the steel stays are to be preferred, provided that they are threaded through both plates and nutted on the fire side. The nuts used should be of stainless steel to make a first-class job and they should also be made thick enough to overlap the end of the stay itself so as to protect it from the direct action of the fire, or they may be of the cap type.

Longitudinal stays in steel boilers are generally fitted by means of steel nipples threaded internally to match the stay and externally to suit the boiler plate. The threads used should not be too fine; 32 t.p.i. is suitable for plates up to $\frac{3}{16}$ in. thick while 26 t.p.i. is better for plates from $\frac{1}{4}$ in. to $\frac{3}{8}$ in. thick.

Securing Fittings to Steel Boilers

Fittings should not be screwed directly into steel boilers unless the plate thickness amounts to $\frac{1}{4}$ in. or so—an unlikely dimension in any but large model locomotive boilers. Pads are generally used, riveted to the outside of the plate or the surface of the barrel as the case may be, to increase the thickness for the threads. These pads are normally used for safety valves, check valves, regulator stuffing boxes or flanges, turret or manifold fittings, water gauges and so on.

Washing Out Facilities

Care must be taken to provide adequate washing out facilities in the larger model steel boilers. The firebox crown, the water legs of the firebox and the bottom of the barrel are the most vulnerable points as regards scale formation. A "manhole" may be provided in the bottom of the barrel near the throatplate in steel boilers of $1\frac{1}{2}$ in. scale and above; the backhead and firebox outer wrapper should also be provided with washout plugs and blowdown valves may be fitted in each side of the firebox at the lowest possible point just above the level of the foundation ring.

CHAPTER SEVEN

Waters, Fuels and Combustion

Feed Water

ALL the boilers so far described have been designed for the normal method of coal firing, which is much the most popular among the builder of small locomotives today. We must now, therefore, turn our attention to some other forms of heating, such as methylated spirit firing, firing by paraffin or petrol blowlamp, firing by propane gas and other methods. But before we do so, a few words on feed waters may be of interest to the reader.

Pure water is a chemical combination of hydrogen and oxygen, and steam, of course, is the vapour of water generated by the application of heat. The process of production of steam on the surface of the water is normally termed evaporation, the constituents of the water and steam remaining the same by volume.

Absolutely pure water can seldom be obtained for model locomotive running, as there are always impurities present, varying in quantity and character, according to the locality.

Water used for locomotive purposes may be classed as follows:

(a) Surface drainage,
(b) River or stream water,
(c) Well water,
(d) Domestic tap water, and
(e) Distilled water.

The last named is clearly the best for model boiler work. Water from surface drainage is probably the next best for model boilers, as its impurities have been absorbed if the district is one of limestone or chalk, where the impurities settle in the form of scale. This can be removed by washing out frequently and periodically de-scaling.

It is of the utmost importance that this de-scaling process should be done fairly frequently and thoroughly, otherwise the water is prevented from coming into close contact with the heating surfaces of the tubes and plates, with the possibility of them becoming burnt. If these deposits are allowed to remain, the effective steam raising capacity of the boiler is greatly impaired. It has been computed in full-size boilers that a layer of scale $\frac{1}{8}$ in. thick could cause a loss of steam raising capacity of no less than 25 per cent.

River water is not generally so suitable as surface drained water, as the characteristics of the impurities are influenced by the gathering grounds of the districts through which the water flows. Many works, particularly chemical or iron works, contribute a great deal of matter detrimental to the water. Most rivers also contain a large amount of suspect matter, which may be of acid character or organic substance. Organic matter, sewage, etc., are often freqent causes of priming.

Little need be said about well water. Because of their varied environments, it would be difficult to classify any well waters in a manner which would assist this present study. It is true that well water often contains more foreign matter than even some river and stream waters.

Treatment of Feed Water

The treatment of raw water to make it suitable for feeding the model boiler is a matter that can only be decided upon after a proper chemical analysis of the water concerned. Expert advice on the subject can, however, be obtained from severl reputable firms who specialise in water softening; there are, nevertheless, certain points that those operating boilers should know about, including a number of simple tests which can be carried out in the amateur workshop.

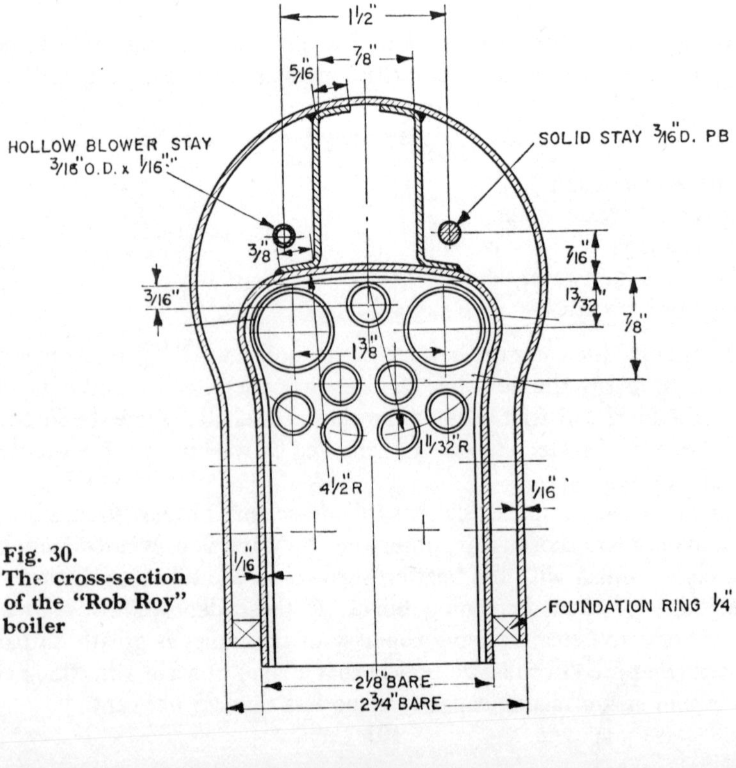

Fig. 30. The cross-section of the "Rob Roy" boiler

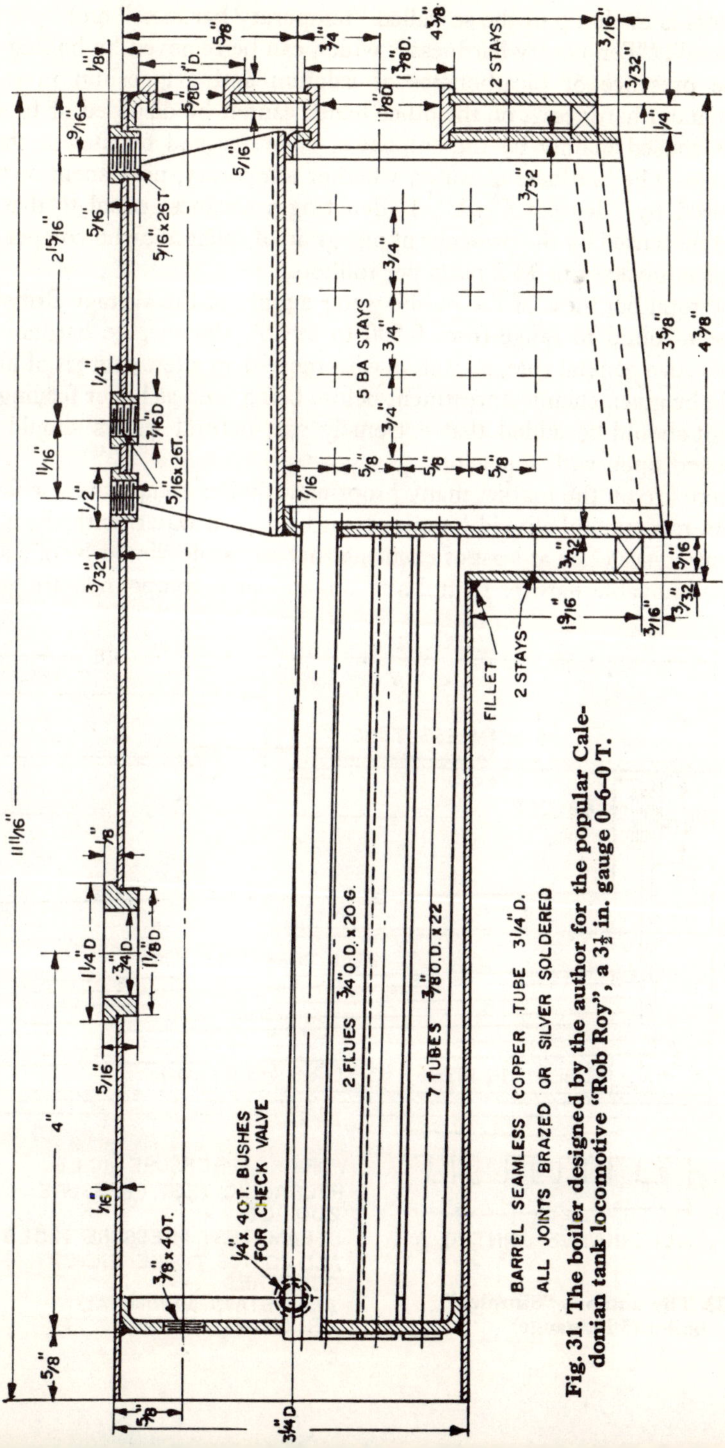

Fig. 31. The boiler designed by the author for the popular Caledonian tank locomotive "Rob Roy", a 3½ in. gauge 0–6–0 T.

The scale-forming properties of water are due to its hardness. The total hardness is made up of the so-called "temporary hardness" and "permanent hardness". "Temporary hardness", which can be removed by boiling, is due to the presence of bicarbonates of calcium and magnesium in solution. "Permanent hardness", on the other hand, cannot be disposed of by boiling and is caused mainly by the sulphates, chlorides and nitrates of the same elements. The hardness of water, whether temporary, permanent or total, is measured by "degrees Clark", 1° denoting a hardness equal to that which would be caused by the presence of one grain of calcium carbonate per gallon. This is equivalent to 14·2 parts per million.

The total hardness of the public water supply of the average British town has been found to range from 0·800 to 39·18°, the average hardness being 13·20°. As a general rule, all waters with more than a few degrees of hardness should be given chemical treatment before being used as boiler feeding water, while it should be added that extremely soft natural waters should always be looked upon with suspicion.

There are on the market many proprietary boiler compounds, which their makers recommend should be mixed with the feed water, with the object of preventing scale; or at least of ensuring that any scale should be of a soft and easily removable nature. In full-size boilers, such compounds are generally

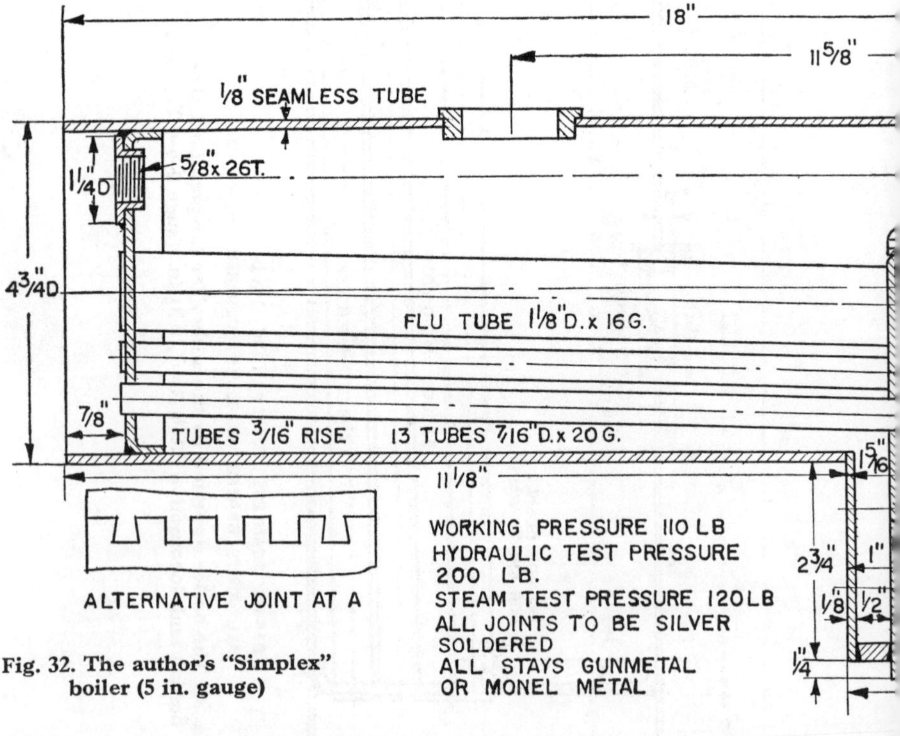

Fig. 32. The author's "Simplex" boiler (5 in. gauge)

dribbled at a constant rate into the feed water in a heater of an open type, on the assumption that the approximately correct amount will be supplied.

These boiler compounds may differ greatly in composition, their principal ingredients, apart from water, being usually soda ash, caustic soda, tri-sodium phosphate and often some organic material, such as tannin, which has the property of causing the lime salts to be precipitated as a soft sludge. When water requires softening, it is generally more convenient to do it outside the boiler by some form of apparatus that can be relied upon to dose the water with the required chemicals in the exact quantities required to give the correct results. The feeding of chemicals into the boiler may cause foaming and priming, as well as involving excessive blowing down.

The simplest way of testing water for hardness is to shake up a sample of the water with increasing quantities of standard soap solution and note the quantity of the soap solution required to form a permanent lather. In tests of raw water, this gives the total hardness. The standard soap solution which is an alcoholic solution of potassium oleate, should be obtainable from any firm of manufacturing chemists.

An old method of conducting the test is to place 70 ml. of water to be tested in a 100 ml. glass-stoppered bottle, into which the soap solution is run, not more than 1 ml. at a time, from a graduated burette.

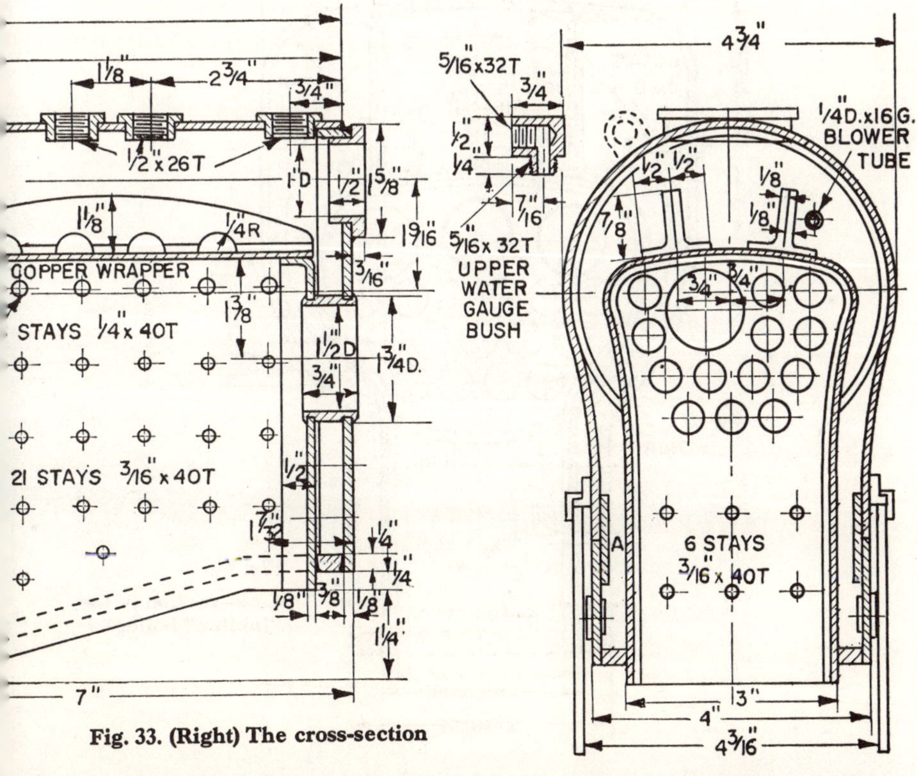

Fig. 33. (Right) The cross-section

The stopper is replaced after each dose and the bottle then violently shaken. The total hardness of the water in degrees Clark is taken as one less than the number of ml. of soap solution that will form a thick lather, lasting for at least three minutes, when the bottle is laid on its side. If the lather does not last as long as this, it is due to the presence of magnesium salts and is to be neglected. The reason for taking a 70 ml. sample is that it contains just as many milligrammes as there are grains in a gallon, while the 1 ml. from the burette is deducted to allow for the quantity of soap required to make a lather with pure water. Waters having a hardness of more than 14° should be diluted with distilled water before making the test, the sample test being 70 ml. as before. If the dilution is half and half, the figure for hardness must be multiplied by two and in proportion for other strengths. In making the soap test it is important that the hands should be clean and the fingers must not be used to close the bottle when shaking it, for the smallest contamination of the sample with grease will destroy the value of the test.

In the above procedure, we have assumed that the hardness is proportional to the quantity of soap used and this is not particularly accurate. Another method can be used.:

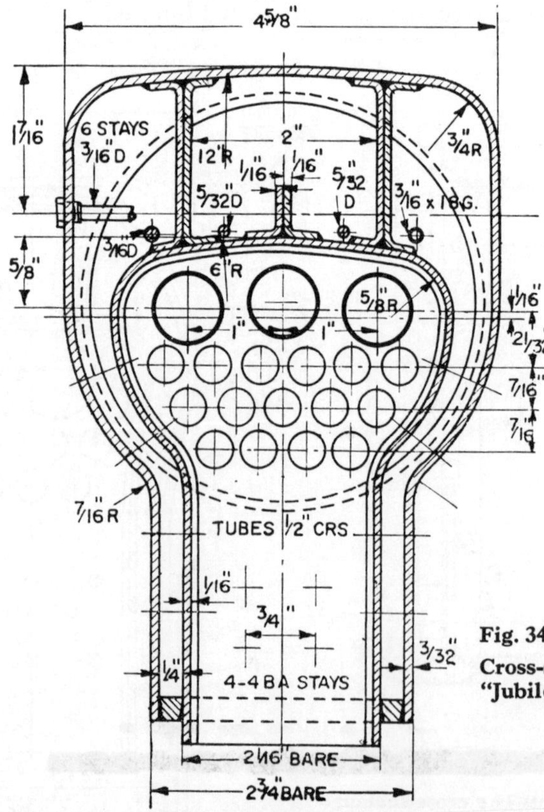

Fig. 34.
Cross-section of the "Jubilee" boiler

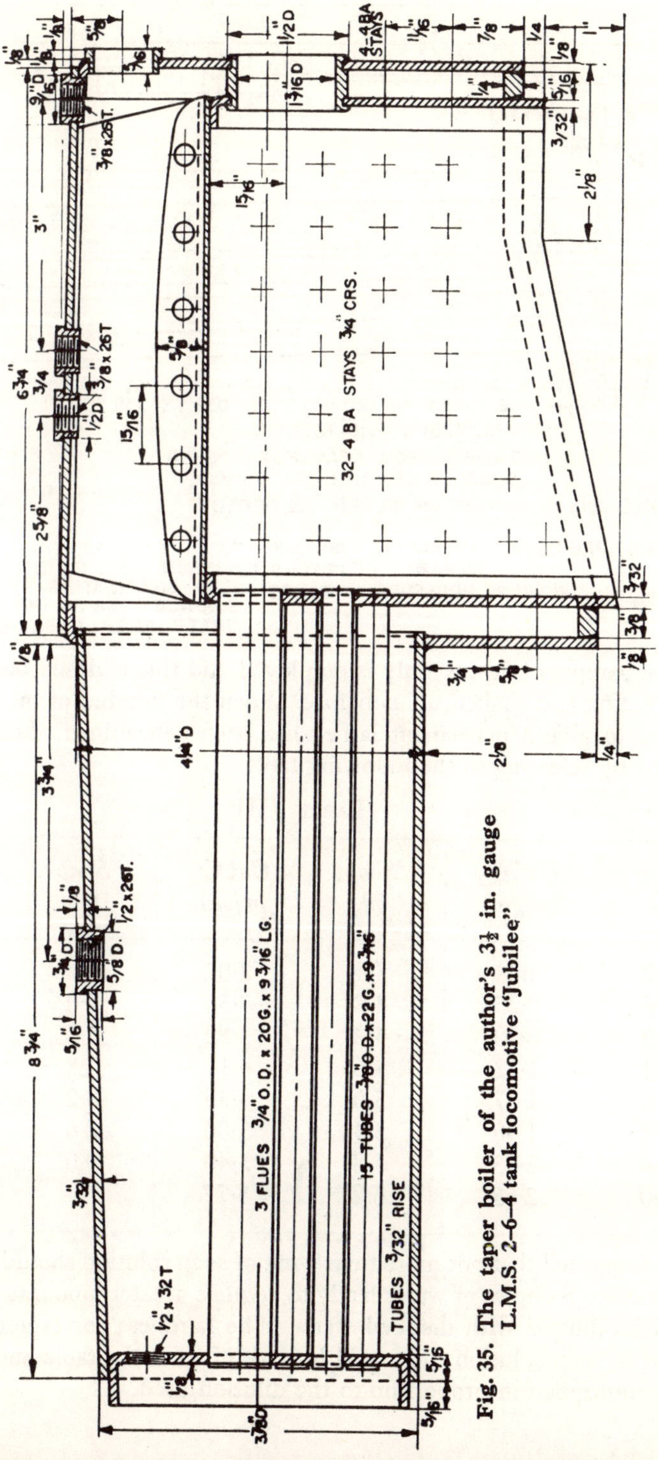

Fig. 35. The taper boiler of the author's 3½ in. gauge L.M.S. 2–6–4 tank locomotive "Jubilee"

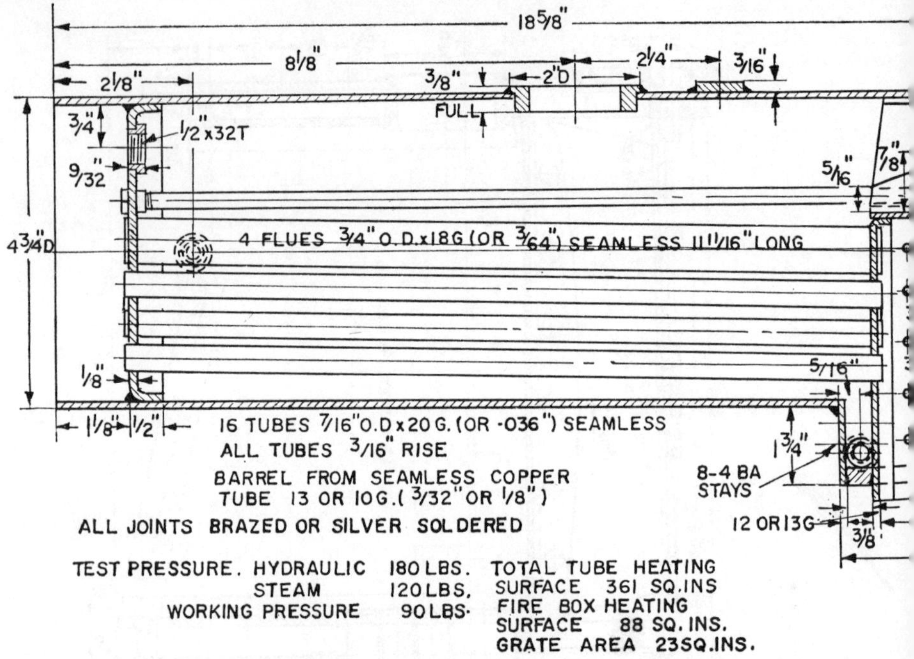

A test sample of 50 ml. only is employed and this is dosed with standard soap solution and shaken up as before. When the number of ml. of solution required to give a permanent lather has been determined, the hardness is obtained by referring to the following table:

TABLE VIII

Soap ml.	*$CaCO_2$ gr/gal.*	*Soap ml.*	*$CaCO_2$ gr/gal.*	*Soap ml.*	*$CaCO_2$ gr/gal.*
0·2	nil.	3·4	3·07	6·6	6·15
0·6	0·25	3·8	3·50	7·0	6·50
1·0	0·70	4·2	3·82	7·4	6·87
1·4	1·12	4·6	4·17	7·8	7·28
1·8	1·47	5·0	4·55	8·2	7·65
2·2	1·90	5·4	5·00	8·6	7·97
2·6	2·25	5·8	5·40	9·0	8·30
3·0	2·65	6·2	5·77		

It is suggested that not more than 9 ml. of soap solution should be used for one titration. Samples of water likely to require greater amounts than 9 ml., should be diluted with distilled water. The hardness corresponding to the amount of soap solution then used is read off from the table and this figure is then multiplied in proportion to the dilution used.

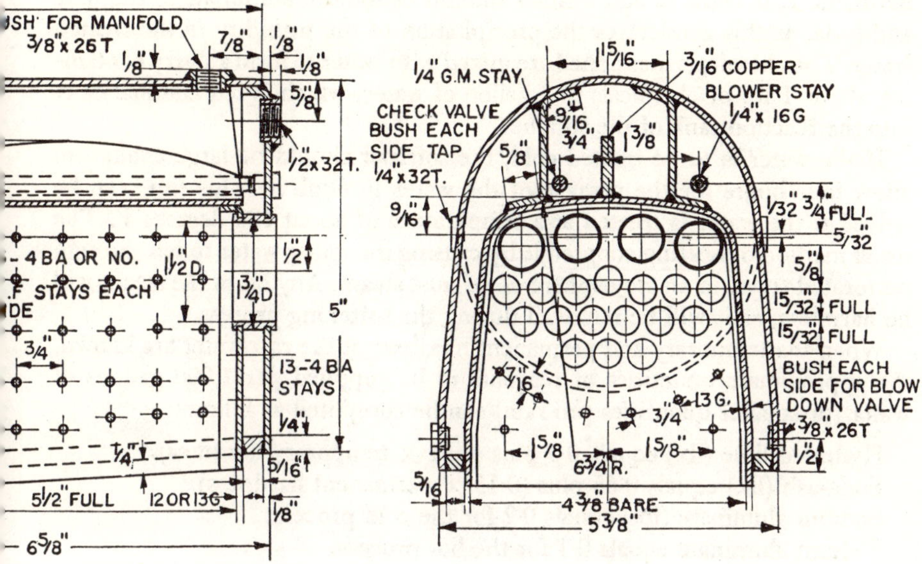

Fig. 36. A wide firebox boiler for a Canadian "switcher" (Caribou)

Temporary and Permanent Hardness

To find out how much of the total hardness of the water is permanent, the following procedure may be adopted. A sample of about 100 ml. of the water is boiled until more than half of it has been evaporated. The remainder is cooled and filtered and is then made up to its original volume by the addition of distilled water. A 50 ml. sample of the mixture is then tested for hardness with standard soap solution, as described previously. The resulting figure will be the permanent hardness of the original water and its deduction from the previously calculated total hardness will be the temporary hardness that was removed by boiling.

Another method of determining temporary hardness is to take 70 ml. of the water, add a few drops of methyl orange and then drop in N/10 nitric acid drop by drop from a burette, until the colour just changes from yellow to pink. Five times the number of ml. of acid required is equal to the temporary hardness in grains per gallon.

Softening Processes

A common method of softening water is by the lime and soda process. The treatment is generally carried out in an automatic water softener, which adds a properly adjusted mixture of the reagents in proportion to the quantity of the water passing through the apparatus.

Hydrated lime, containing about 97 per cent of $Ca(OH)_2$, is usually employed to remove the temporary hardness and sodium carbonate (soda-ash), containing practically 100 per cent of Na_2CO_3, removes the permanent hardness. It is usual to add a small amount of sodium aluminate to the lime and soda, as this accelerates the precipitation of the turbidity in the treated water. The chemicals mentioned are mixed with water to form a slurry containing about 1 lb. of chemicals to 1 gallon of water, which is fed automatically into the reaction tank of the softener.

If the water is to be treated cold, the softener should be large enough to allow four hours for the passage of the water through it. This time can be halved if the water is treated at a temperature of about 180 degrees F. The usual method of heating adopted is by causing incoming water to fall through perforated trays, in an atmosphere of exhaust steam. Any oil in the steam will be harmless, as it will be removed during the softening process.

When the temporary and permanent hardness of the raw water are known, the approximate quantities of reagents to be supplied per 1,000 gallons of water (or smaller quantities pro rata) can be computed as follows:

Hydrated lime (lb.) equals 0·3 plus (0·11 × temporary hardness).
Soda-ash (lb.) equals 0·65 plus (0·15 × permanent hardness).
Sodium aluminate (lb.) equals 0·2 for the cold process.
Sodium aluminate equals 0·1 for the hot process.

The above figures include a small excess of lime and soda, to ensure that the reactions will be as complete as possible.

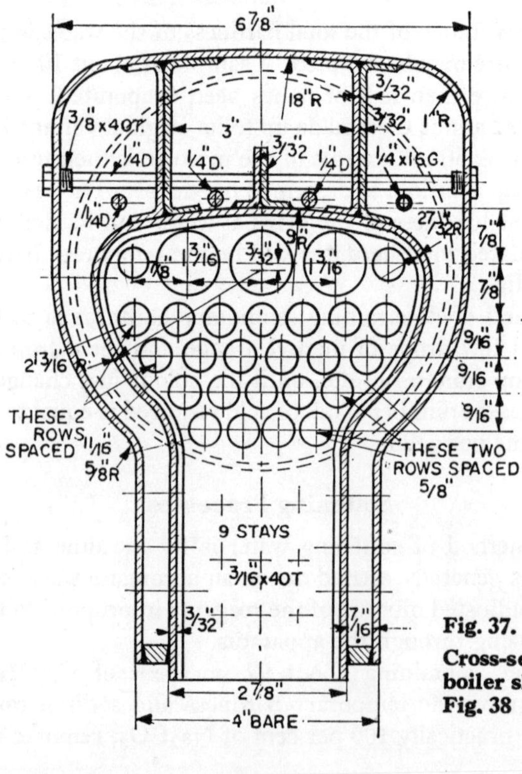

Fig. 37.
Cross-section of the boiler shown in Fig. 38

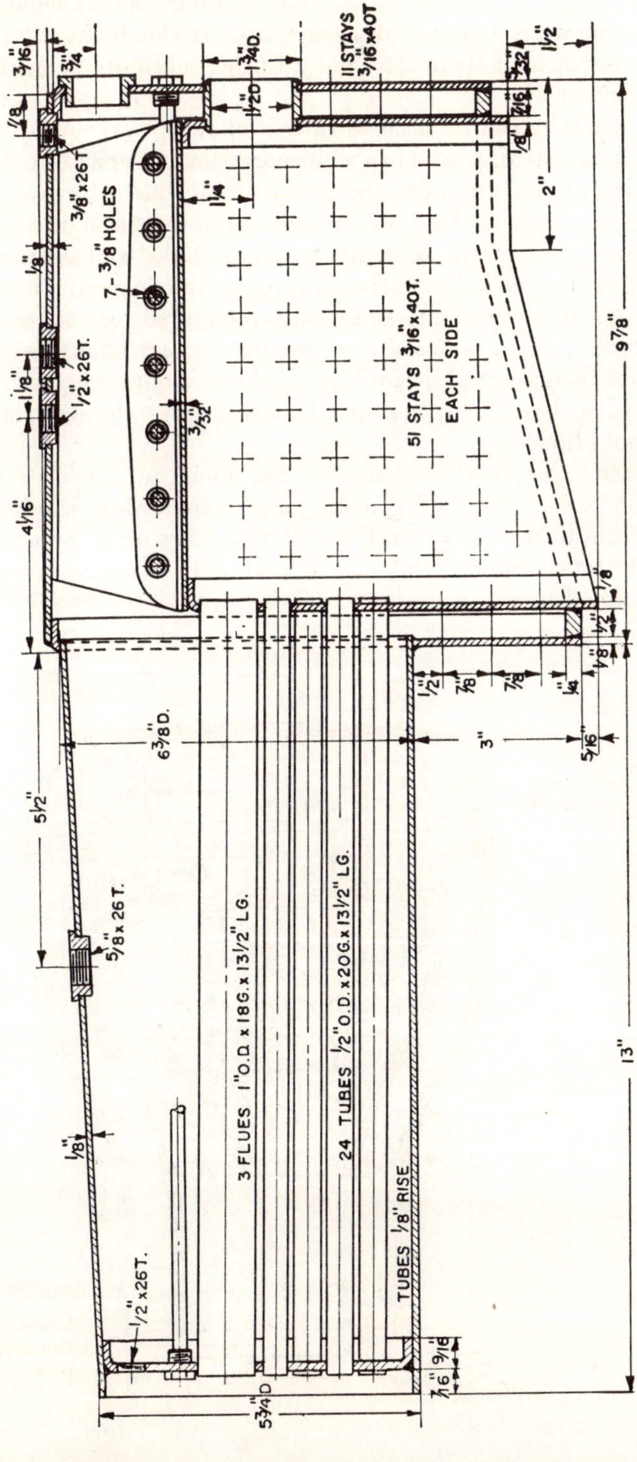

Fig. 38. A boiler suitable for a 5 in. gauge L.M.S. 2–6–4 tank locomotive

Water properly softened by the lime and soda process should have a total hardness between 1 and 2 degrees Clark. It should have an alkalinity to phenol-phthalein of about 2·5° and a further alkalinity to methyl orange of about another 2°.

The alkalinity can be decided in the following way: A 70 ml. sample of the softened water is placed in a white porcelain dish and two drops of phenol-phthalein indicator are added and gently stirred with a glass rod. The indicator is made by dissolving 0·2 g. of pure phenol-phthalein in 60 ml. of colourless methylated spirit. 40 ml. of distilled water is added and the mixture filtered. The addition of the indicator should turn the sample pink; if it does not do so, it is a sign that more lime and soda are required for softening. When the water turns pink, it is dosed drop by drop from a burette with decinormal nitric acid, being stirred meanwhile, until the colour disappears. The number of ml. of acid required, multiplied by five, gives the degrees of alkalinity to phenol-phthalein.

The same sample, now colourless, has added to it two drops of methyl orange indicator, which will give it a yellow colour. More acid is then dropped in with continual stirring, until the solution takes on an orange colour. Five times the total number of ml. of acid used in the two titrations gives the total degrees of alkalinity.

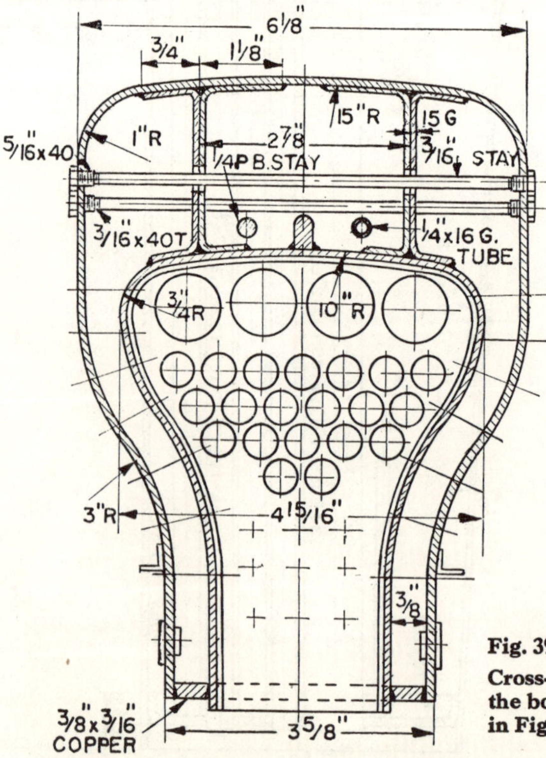

Fig. 39.
Cross-section of the boiler shown in Fig. 40

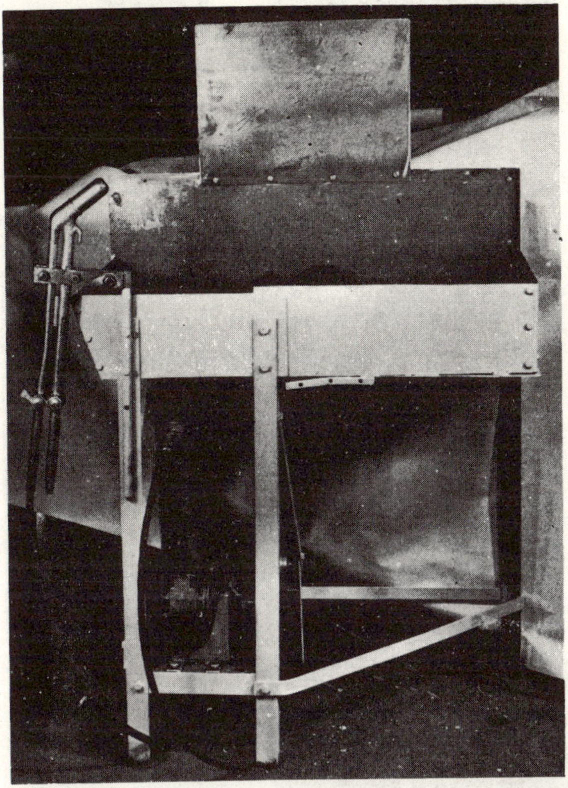

A brazing hearth equipped with air blower, and air/gas blowpipe

Mr. E. C. Dearman and assistant silver soldering the tubes in his "Springbok" boiler

A large Sievert propane blowpipe in use on a big wide-firebox boiler: Photograph courtesy William A. Meyer Ltd

This picture shows the sliding type of firedoor on a 5 in. gauge Great Western—Vale of Rereidol tank locomotive

Plate 34

Mr. E. C. Dearman in Australia gets help from his wife with small blowlamp when fixing tubes to his "Springbok" firebox

Right: Boring holes in smokebox tubeplate for the superheater flues

Below: The 5 in. gauge "Firefly" built jointly by the author and Norman Spink

Some of the equipment used in his tests of his locomotive boiler by Mr. J. Ewins

Left: A hydraulic test on a small water tube boiler

Below: Further views of Mr. J. Ewins' test equipment

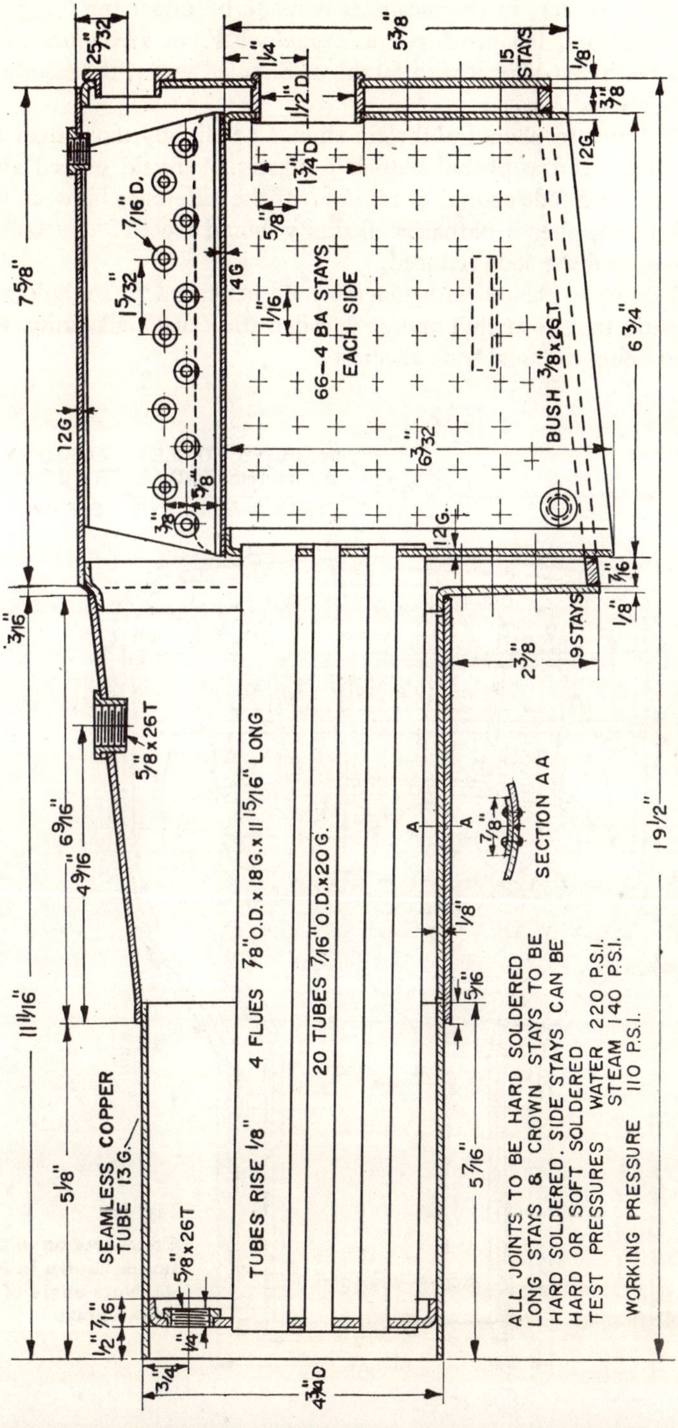

Fig. 40. The boiler for the author's 5 in. gauge Great Western 45xx class 2–6–2 tank engine "Firefly"

The methyl orange indicator (like the phenol-phthalein indicator) can be obtained from dealers in chemicals. It is made by dissolving 0·2 g. of pure methyl orange and 0·5 g. of xylene cyanole F.F. in a mixture of 50 ml. colourless methylated spirit with 50 ml. of distilled water, the solution being afterwards filtered.

The alkalinity to phenol-phthalein should be slightly more than half the total alkalinity. If the phenol-phthalein alkalinity should exceed about 4°, both the lime and soda should be reduced. If the difference between the total alkalinity and the phenol-phthalein alkalinity should exceed 2°, the lime should be increased and the soda reduced.

The alkalinity to phenol-phthalein is sometimes called the sodium carbonate alkalinity and that to methyl orange is called the caustic alkalinity, the sum of the two being the total soda alkalinity.

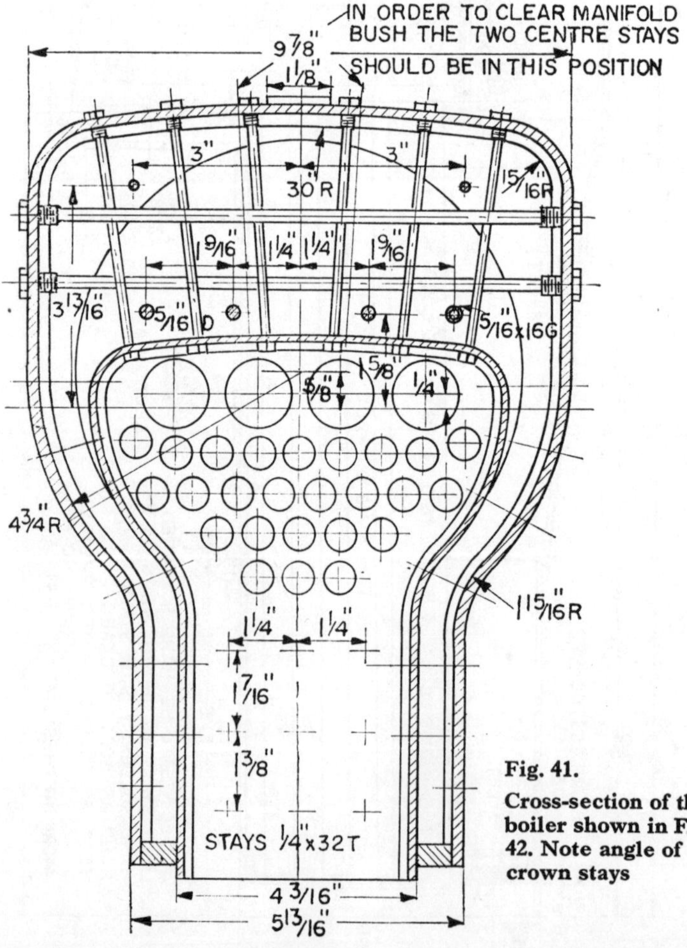

Fig. 41.

Cross-section of the boiler shown in Fig. 42. Note angle of crown stays

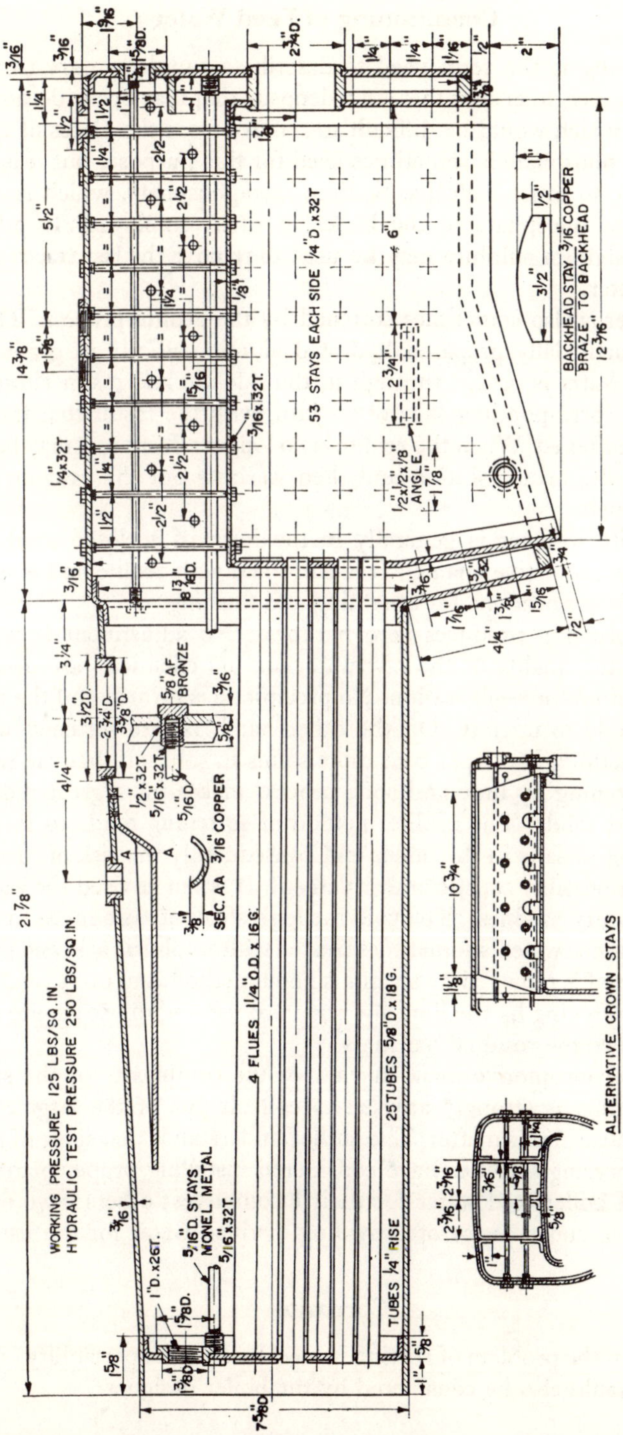

Fig. 42. The boiler for the author's $7\frac{1}{4}$ in. gauge L.M.S. standard class five 4–6–0 "Highlander"

Conditioning of Feed Water

Conditioning is the term used to describe a supplementary treatment of boiler feed water, to ensure that such deposits that may occur do not produce a hard scale which would be difficult to remove from the plates of the boiler. Tri-sodium phosphate is sometimes used for this purpose, but it is not ideal as it is liable to cause deposits. Sodium metaphosphate, which is converted to tri-sodium phosphate in the boiler, is often employed. In addition to phosphate, sodium sulphite may be used to remove the last traces of oxygen from the water.

Feed water is also sometimes softened by the zeolite process. The zeolite used is an artificially prepared hydrated alumino-silicate in granular form. When hard water is passed through it, the calcium and magnesium salts are replaced by corresponding salts of sodium, with the result that the water is completely softened. When the zeolite is no longer effective, it may be restored by flushing the brine solution and then washing out the calcium chloride, thereby formed.

The zeolite softener is generally in the form of a closed steel cylinder, finished with protective enamel. It is half filled with zeolite and is fitted with the necessary connections for operating and flushing.

Zeolite softening produces zero hardness. No adjustment is needed for variation in the quality of the raw water and no chemicals are used, except the salt required for regeneration. No precipitate is formed and the apparatus is quite simple to operate. On the other hand, zeolite softened water will contain something like 5 per cent more solids in solution that the raw water, whereas softening by lime and soda process makes for a great reduction in the dissolved solids. The zeolite method of softening tends to increase the concentration of salts in the boiler and consequently the risk of foaming and priming and possibly caustic embrittlement. It is not advised for waters with much temporary hardness. To avoid the high alkalinity produced in the latter case, the softened water is sometimes treated with sulphuric acid and the carbon dioxide formed is removed by steam. A good method is to carry out the major part of the softening by the lime and soda process and to use zeolite softening only to remove the residual hardness.

The model locomotive man, in view of the complexity of the subject of water softening, is strongly advised to consult one of the large companies which specialise in this matter. The author understands that some of the larger model engineering societies have considered installing proper plants for the treatment of boiler feed water for their locomotives; others have even gone to the trouble and expense of providing distilled water for the use of their members.

Corrosion

Apart from the problem of rusting in steel boilers, the possibility of galvanic corrosion should also be considered by the boiler designer.

Galvanic corrosion may be defined as the accelerated electro-chemical corrosion produced when one metal is in electrical contact with a more "noble" metal, both being immersed in the same corroding medium which is called an electrolyte. The metal that does not corrode is called the noble metal. During the corrosion process, an electric current is generated by the two metals in contact, and the magnitude of the current is equivalent to the acceleration of the corrosion of the more vulnerable (less noble) material beyond its normal rate. The general reaction is similar to that on a miniature battery from which a finite amount of electrical power is derived. In a wet cell, for example, the less noble zinc suffers accelerated corrosion while the copper is almost unattacked.

The coupling of two metals of differing "nobility" will result in galvanic or accelerated corrosion of the less noble metal. The actual galvanic effect is determined by the quantity of the current rather than the potential difference.

To minimize galvanic corrosion, metal combinations should be avoided where the area of the less noble metal is relatively small. It is good practice to use the more noble metals for the smaller parts in the model boiler—i.e. copper tubes in steel boilers.

Silver-soldered joints are always better than riveted or screwed joints from the point of view of galvanic corrosion.

In large model steel boilers, using copper tubes, sacrificial zinc plates may be fitted. These may be in the form of tubes attached to the insides of the wash-out plugs—for easy access.

Chloride corrosion has also been observed when tap water, which has been treated with chlorine for domestic purposes, has been used in boilers made of certain types of stainless steels. This chloride corrosion is a slow, but accumulative process, but insufficient data is available at the present time, to allow the author to give definite advice on the subject.

Coals

As mentioned previously, coal is much the most popular fuel for the small steam locomotive of today. Coal is derived from the remains of ancient forests which became buried under accumulations of silt and were subjected to great pressures and high temperatures for thousands of years. The timbers of these ancient forests underwent various physical and chemical changes, thus giving us the various types of coal which we know today, varying from lignite or brown coal to the best quality anthracite, which contains a very large proportion of carbon.

The principle changes which took place in the transformation from wood to coal was an increase in the percentage of carbon and a reduction in the oxygen, hydrogen and nitrogen contents. Thus wood contains approximately 48 per cent of carbon and 40 per cent of oxygen, whereas the best Welsh anthracite contains about 92 per cent of carbon and about 1·6 per cent of oxygen.

In the author's opinion, the best solid fuel for use in model locomotive boilers is a mixture of the best Welsh steam coal and the best anthracite, about two-thirds of the former being used and one-third of the latter. Such a fuel will have a calorific value of about 15·3 B.T.U. per pound. However, quite good results are obtained by using steam coal only and good results have also been obtained from some of the "patent" fuels obtainable from the coal merchants.

When the better qualities of steam coal or anthracite have been unobtainable, some of the bituminous coals have been tried and will raise steam, but their heavy smoking qualities and the formations of residues in the tubes and flues will discourage their use except in emergencies.

To obtain the best results in burning solid fuel in model locomotive fireboxes, the coal should be broken up to a sensible size and the dust sifted out before use. Reasonable sizes of lumps are as follows: for ½ in. scale locomotive boilers—about the size of peas; for ¾ in. scale boilers—about the size of broad beans; and for 1 in. scale locomotive boilers—about the size of walnuts, larger sizes in proportion.

Combustion of Coal

Coal is said to burn when it unites with oxygen, giving out heat. The oxygen is drawn in, in the air through the grate, the air being a mixture of about 21 per cent oxygen and 78 per cent of nitrogen by weight and 1 per cent of water vapour so that for every 5 cu. ft. of air passing through the firebox, about 4 cu. ft. contains nitrogen and 1 cu. ft. of oxygen. The nitrogen is an inert gas chemically and does not assist in the burning of coal. The nitrogen. thus retards the rapidity and consequently the temperature of combustion. There are other substances besides carbon in coal and these include oxygen, nitrogen, hydrogen, sulphur and water.

When coal is burned, part of the hydrogen in the coal is united with the oxygen forming water vapour or steam. The excess hydrogen unites with some of the carbon at around 800 deg. F. giving off hydro-carbon vapour. When the engine is in steam and the firebox temperature has risen to about 2,500 deg. F., these hydro-carbons unite with the oxygen in the air and burn with an almost smokeless flame. Carbon dioxide is then formed and this is the product of good combustion. If there is insufficient oxygen, i.e. insufficient air passing through the fire, some of the hydro-carbons escape unburned and black smoke is produced. Such gases are termed carbon-monoxide and are the products of poor combustion. The hydrogen in the coal and the oxygen in the air give out an intense heat on combining with the carbon, thus producing the incandescent masses of coal in the firebox. The sulphur in the coal causes it to clinker and adhere to the firebars, thus preventing a free flow of air from the ashpan through the grate.

The nitrogen in the coal is only a small proportion and takes no part in this combustion, passing up the chimney taking some of the heat with it. On average, 1 lb. of coal takes about 12 lb. of air to burn it, of which over 9 lb. are nitrogen. The nitrogen reduces the rate of combustion as it has to be heated up to the temperature of the firebox and thus causes a great deal of heat to be wasted. It will be understood from this that if too much air is admitted to the firebox, there will be a loss of heat. If black smoke is seen from the chimney, extra air will probably be required and some of this can be provided by either leaving the firehole door slightly open or if the door is of the type where additional airholes can be uncovered, these can be opened.

The brick arch, if made of a proper refractory material, assists combustion once it has reached a suitable temperature and it also deflects the heated gases towards the back of the firebox, thus enabling them to mix with the extra air admitted through the fire door. The brick arch in full-size boilers helps to protect the tube ends from the direct action of the fire and a deflector plate is usually provided to deflect the cold air in a downwards direction so as to prevent a direct rush of cold air onto the tubeplate. This, however, is not important in small locomotive boilers built of copper and with silver soldered joints.

Regarding the amount of extra air to be allowed in through the firehole door, this should not be excessive as it all has to be heated up to the temperature of the firebox gases and if it is not so heated up, it will lower the temperature of the gases passing through the tubes, reduce the steam rate of the boiler and reduce the effectiveness of the superheaters. A safe rule is to allow sufficient air to be admitted through the firehole door to ensure that the smoke is completely consumed, but no more. In the operation of model steam locomotives, one often sees the firehole door opened wide to cool the firebox and tubes and thus to prevent excessive blowing off. This is not, however, a good practice and the proper action to take is to reduce the air passing through the fire by closing the dampers which should be provided at the rear end of the ashpan.

Thickness of Fire

As in full-size practice, the fire in the locomotive firebox should not be built up too deeply. In general, it should be somewhat thicker at the sides than in the middle so as to prevent gases creeping up the side of the firebox and being drawn off through the tubes unburnt. The coal may, however, be built up a little higher just underneath the firehole door.

Firegrates

It is sometimes forgotten that when the fire in the firebox is burning really brightly so that it is nearly white hot on the top, the grate itself will also become red hot and this explains why grates made of ordinary mild steel and even those made of cast iron do not last very long. Thus stainless steel may be considered for the firegrate as mentioned in a previous chapter.

Some boilers have a long narrow grate sloping down towards the front of the firebox so that when the locomotive is running, the coal has a tendency to fall gradually from the back to the front of the grate. When firing this type of grate, therefore, too much coal should not be fed to the front end. Some grates are made partly straight and partly sloping towards the front and here too, care should be taken not to build up too deep a fire at the front.

The Blast Pipe

The amount of air which is drawn through the grate and the firehole door depends on the blast, i.e. exhaust steam passing from the blast pipe up the chimney to the atmosphere. This creates a partial vacuum in the smokebox which draws the heated gases from the firebox through the tubes. When the engine is working hard, particularly with a late cut-off, as for instance on a long up-grade, the blast will be heavy and could well raise the smokebox vacuum to 6 in. or 7 in. of water. On the level, working with an early cut-off, and a partly opened regulator, the blast will be light and may produce a smokebox vacuum of about 3 in. of water.

Evaporation of Water

If water is boiled in an open vessel at sea level, the temperature of the water and of the steam rising from it will be 212 deg. F. If water is boiled at a level much higher than sea level, the temperature of the water and steam will be much less than this. Similarly, if it was possible to boil water far below sea level, the boiling point would be much above 212 deg. F. owing to the rise in pressure of the atmosphere.

Gauge pressures are measured above atmospheric pressure and are approximately 14·7 p.s.i. below absolute pressure. Thus 180 lb. gauge pressure is approximately equal to 195 p.s.i. absolute pressure.

As an indication of the rise in temperature of the boiling water in the boiler as the pressure rises, the following figures may be of interest: absolute pressure of 80 p.s.i.—temperature 312 deg. F. Absolute pressure 100 p.s.i.—temperature 327 deg. F. Absolute pressure 180 p.s.i.—temperature 373 deg. F.

Steam in contact with the water from which it is being generated is known as saturated steam and contains minute particles of water. This explains why condensation is such a problem in the cylinders of locomotives which are not fitted with superheaters.

Steam is said to be superheated when it has been heated to a temperature above that for the corresponding pressure of saturated steam. Dry saturated steam theoretically contains no water particles and will be approximately the normal temperature for the corresponding pressure.

CHAPTER EIGHT

Oil and Gas Firing

AS HAS been mentioned previously, coal firing is much the most popular form of heat for model locomotive boilers, but some enthusiasts still prefer other methods such as the use of methylated spirit, paraffin blowlamps, petrol blowlamps and bottled gases.

Methylated Spirit

At one time, methylated spirit burners were used in locomotives as large as $2\frac{1}{2}$ in. and $3\frac{1}{4}$ in. gauge. Today, it is seldom found in gauges larger than $1\frac{3}{4}$ in. Methylated spirit is comparatively expensive, but it is clean and, handled with reasonable intelligence, safe. It gives a clear flame and does not make or deposit soot; it does, however, require plenty of ventilation otherwise it will not burn completely, thus wasting its heating possibilities and producing unpleasant fumes.

There are three principal methods of methylated spirit firing: (1) the simple wick lamp, (2) vaporising lamps, (3) force-feed burners.

The simple wick lamp generally consists of a number of short vertical tubes made of thin brass which are silver soldered to a feeding tube arranged horizontally. The wicks can be made from the material sold for use in commercial and domestic paraffin heaters. They should be cut to such a length that they protrude approximately $\frac{1}{8}$ in. above the tops of the burner tubes and reach to the bottom of the supply tube. The supply tubes should be of large enough diameter to ensure that the burner tubes furthest away from the source of supply are not starved of the spirit. The vaporising lamp is normally provided with a wick-fed pilot flame and the principle is simply that as the methylated spirit is sucked up into the vaporiser, it is heated to such a temperature that the spirit is vaporised (i.e. converted into the gaseous state) and it is then allowed to emerge through a large number of very fine holes to which a flame is applied so that a small flame will burn from each of the holes.

It is important with all methylated spirit lamps that plenty of air is allowed to be drawn in all around the flames and also that the tubes of the burners should not be arranged too close to the underside of the boiler. One trouble which sometimes occurs with methylated spirit burners is that the heat is conducted back to the spirit container, causing the spirit in this to be vaporised

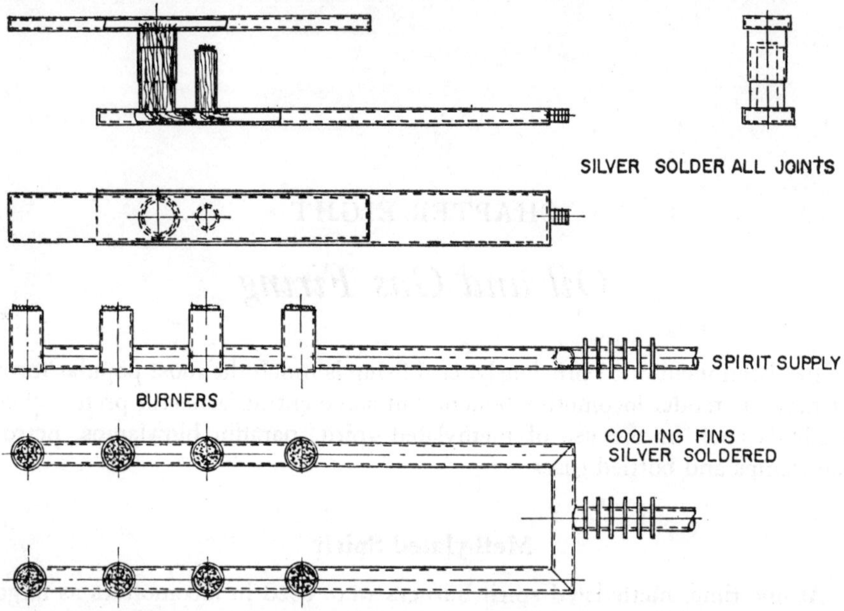

Fig. 43. Two methylated spirit burners. The upper one is of the vaporising type

prematurely. To prevent this, the container should not be placed too close to the burners and cooling fins made of thin gauge material can be silver soldered to the supply tube or tubes in between the container and the burners. All the material used in the construction of the burners and their attendant pipes should be of the lightest gauge metal possible, brass being preferred and all joints should be silver soldered.

Methylated spirit burners on the force-feed principle have been made. The spirit container is made of heavier gauge material so as to stand a considerable pressure. It is provided with a needle control valve and a non-return valve to take a bicycle tyre pump. A pipe is led from the needle valve to the burner and this is coiled around the burner in a similar fashion to a paraffin blowlamp. The feed pipe ends in a very fine nipple, once again similar to that used in paraffin blowlamps. To operate this apparatus, the container is partly filled with spirit and air is pumped into the container to a pressure of about 20 p.s.i. The vaporising tubes around the burner are then lightly heated, the needle valve slightly opened and a flame applied to the jet.

Although the author understands that burners made on this principle have proved successful, he has not had any success himself in experiments and is, therefore, somewhat reluctant to recommend the idea to readers.

Petrol and Paraffin Blowlamps

Petrol and paraffin blowlamps are not much used nowadays for model locomotive work. Petrol is much more easily vaporised than is paraffin. It has

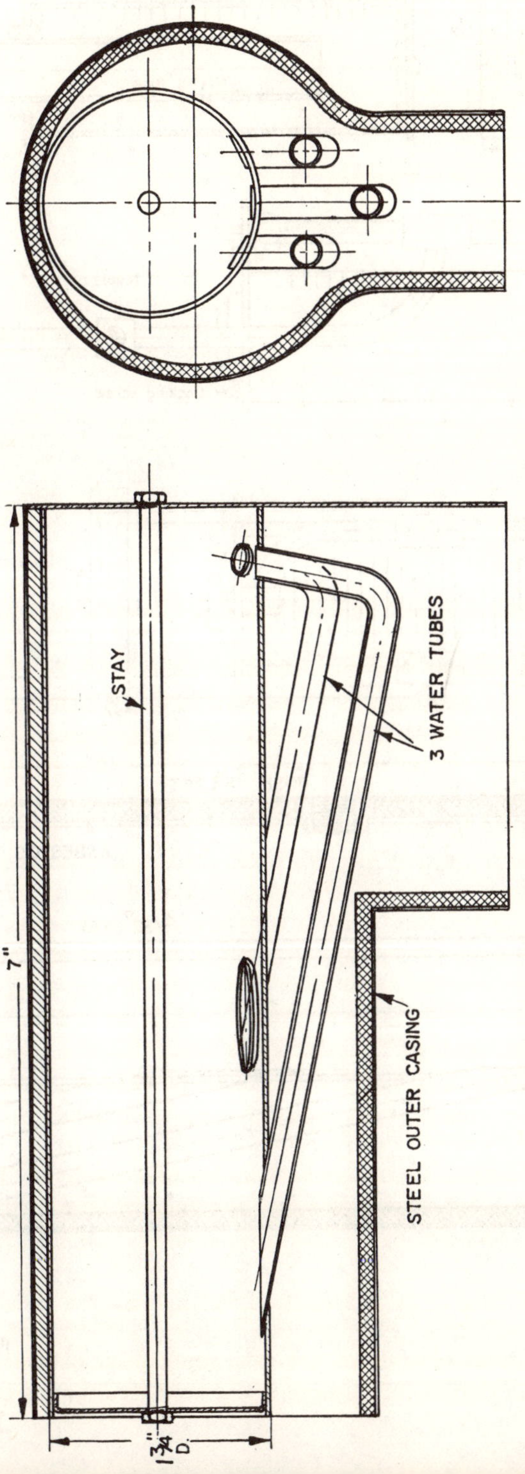

Fig. 44. A design for a simple water-tube locomotive boiler for gauge one, using methylated spirit or propane firing. Note asbestos lagging

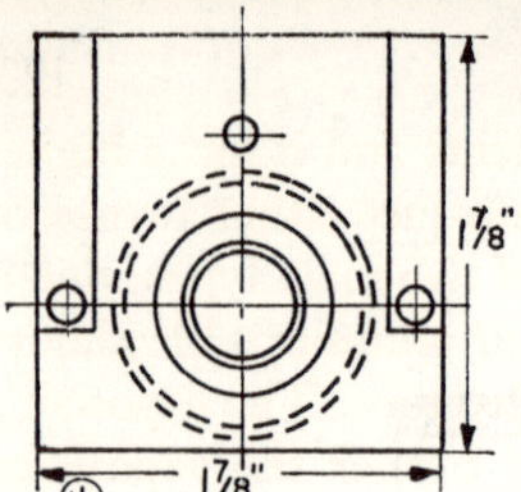
1 7/8"
1 7/8"
①

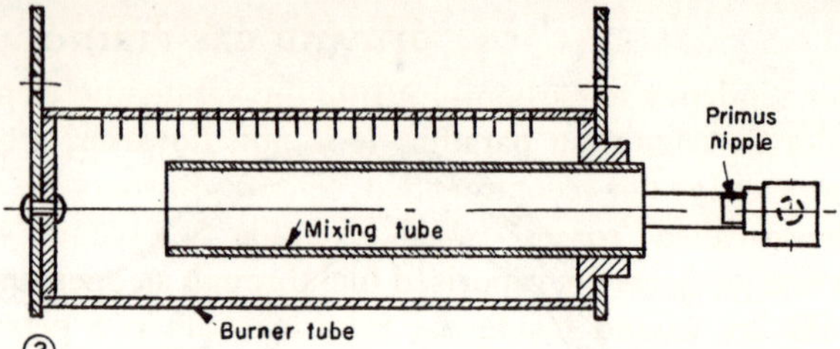
Primus nipple
Mixing tube
Burner tube
②

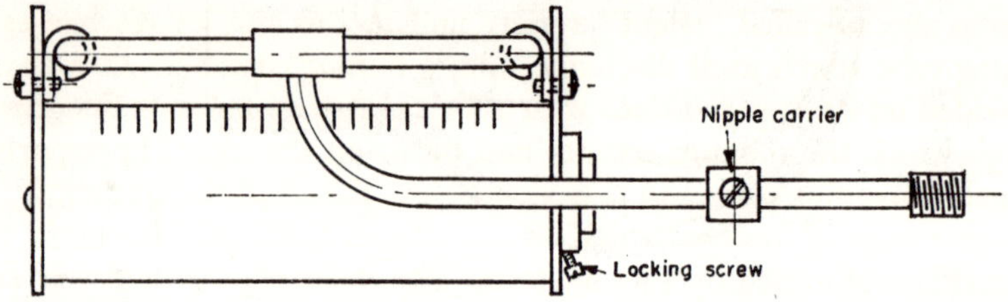
Nipple carrier
Locking screw
③

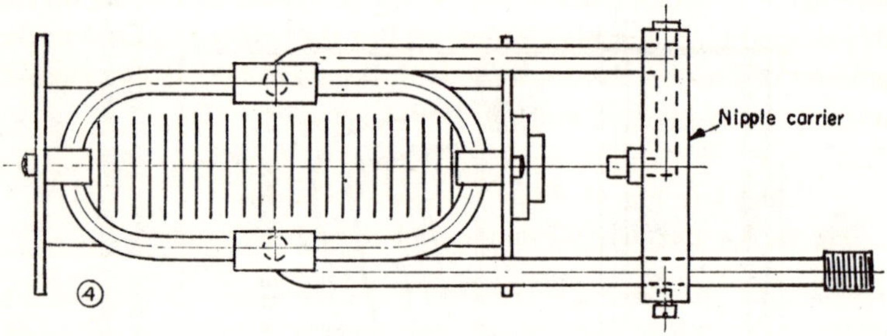
Nipple carrier
④

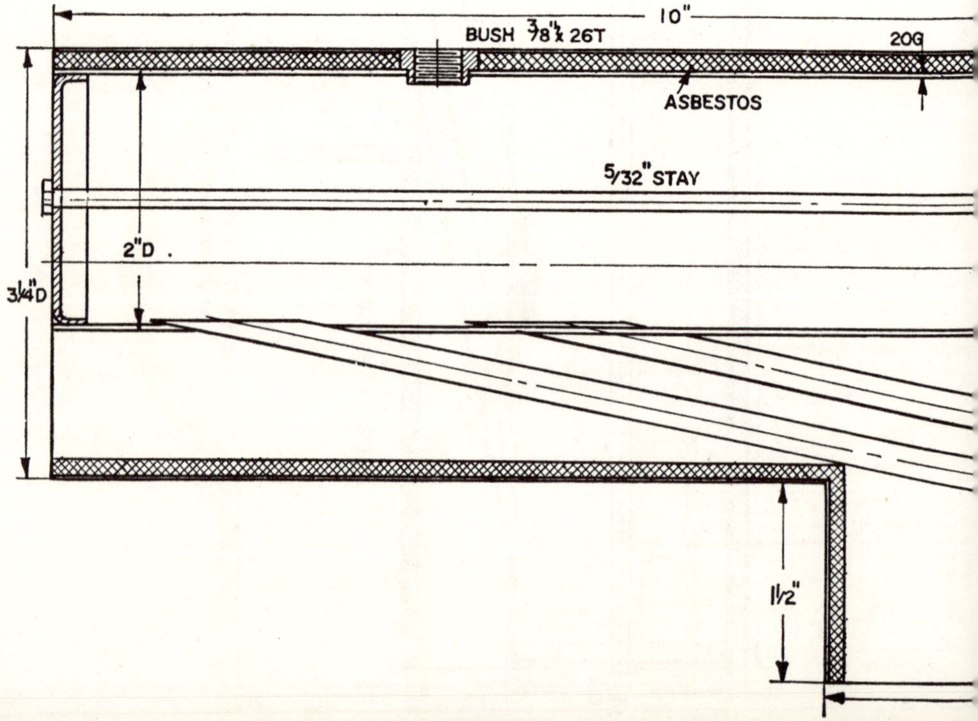
10"
BUSH 3/8" x 26T
20G
ASBESTOS
5/32" STAY
2"D .
3 1/4"D
1 1/2"

less tendency to carbonise inside the vaporising elements and it has a higher calorific value than paraffin. It is not, however, a safe fuel either to use or store and is not recommended. Paraffin burners are of two main types: one is the so-called "roarer" which is similar to a painter's or plumber's blowlamp. A fine jet ejects the vaporised fuel through an open ended tube. The vaporising coils are wound inside the tube whereas in a petrol burner the vaporising coils are generally arranged outside the tube. The second type of paraffin burner is the so-called "silent" variety and in this the jet discharges into a mixing tube which itself discharges into a burner chamber. This chamber is provided on its upper surface with a series of small holes or fine saw cuts through which the mixture escapes and burns on the top. The vaporiser is arranged above this plate and is heated by the flames of the burner.

Another type of burner which has sometimes been used is one in which the paraffin is atomised by a jet of steam. The steam should be dry or superheated and a needle valve should be used to control it. Some of these burners have been made in which the atomised spray is created by forcing oil through a suitably shaped nozzle at high pressure; but the orifice required to pass the small amount of fuel needed for a model locomotive boiler is so small that it becomes very easily clogged and put out of action.

Top left: Fig. 45. Silent type paraffin burner
Fig. 46. A water-tube locomotive boiler for 2½ in. gauge

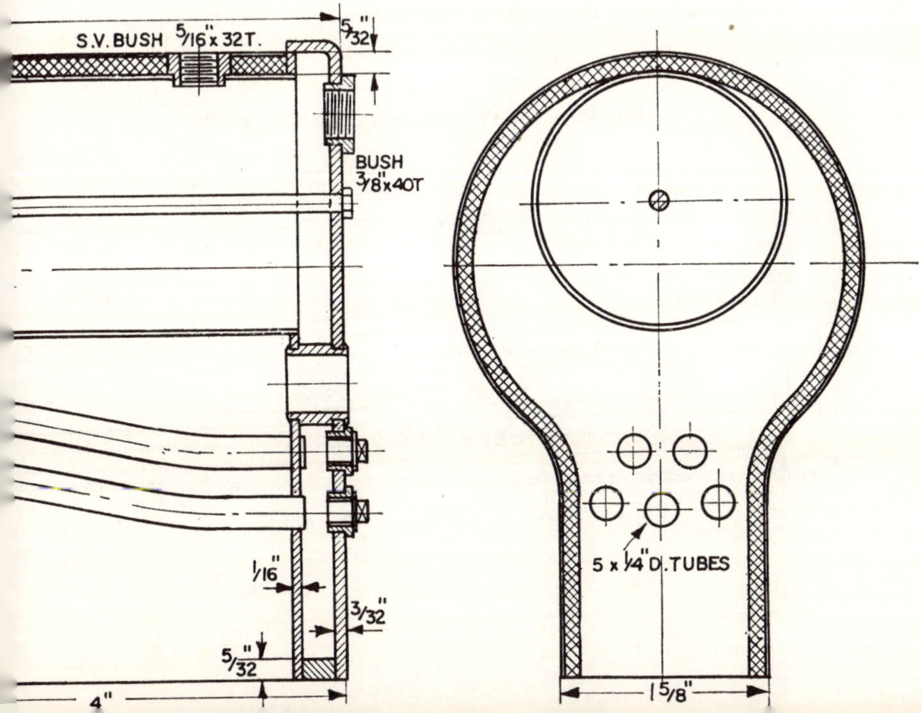

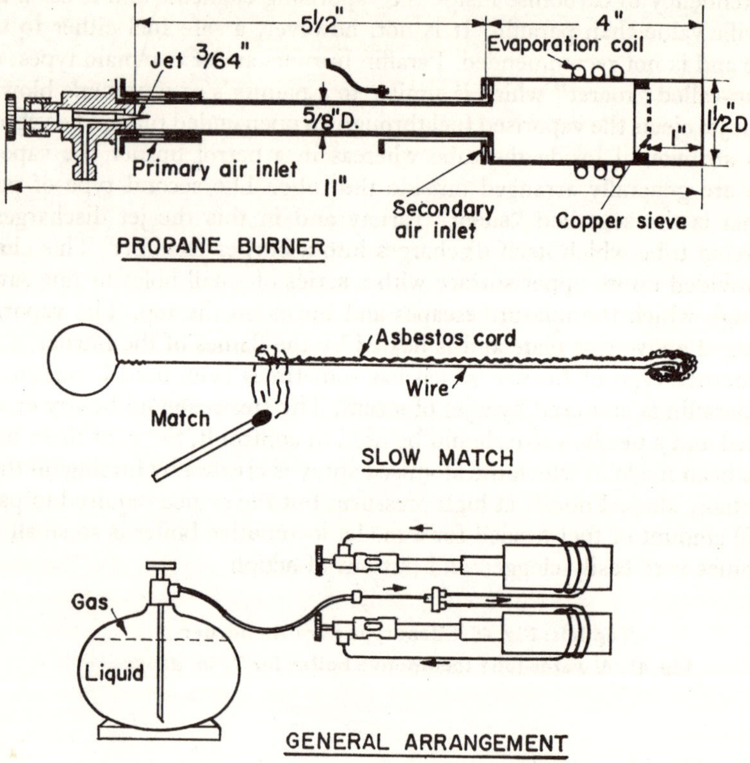

Fig. 47. A propane burner designed by Herr Grosse-Holtfort of Hamburg

Fig. 48. A wide firebox boiler for Gauge "O"

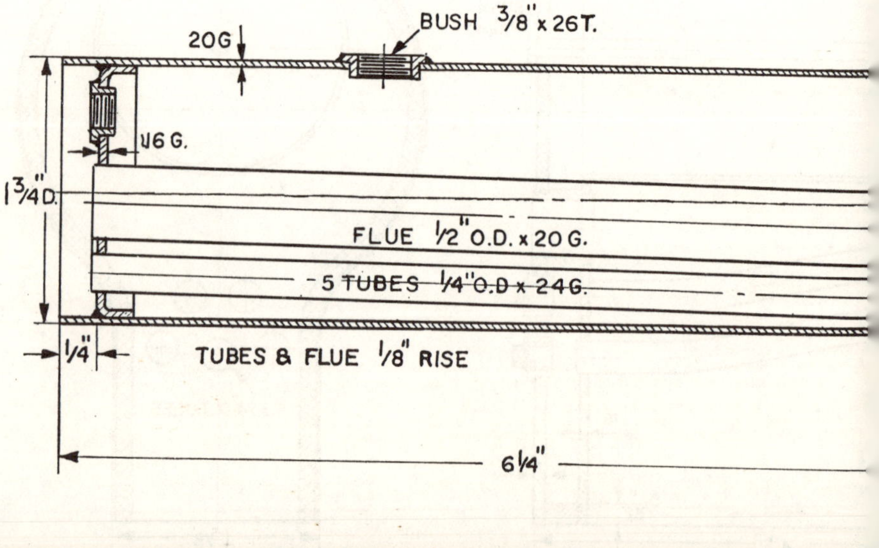

Bottled Gas Burners

Having had some experience of the more recent types of burners and blowpipes designed to work with bottle gas, the author is of the opinion that this method of heating is much to be preferred to methylated spirit, petrol or paraffin.

Bottle gas can be obtained either as butane or as propane. These gases are very similar but as propane has a slightly higher heat value, it is to be preferred of the two.

Propane burners have been used in model steam locomotives as small as gauge "O" ($1\frac{1}{4}$ in. gauge) with every success. The author recently designed a scale model Royal Scot 4-6-0 locomotive in which the boiler was of the normal locomotive firetube type except that the firetubes were made of comparatively small diameter and of thin gauge, and the burner was adapted from one of the propane burners manufactured by the Swedish firm of Sievert. The burner tube was flattened so as to splay the flame out into a fan shape and it was arranged at approximately 45 degrees to the inner firebox and provided with a gas-tight union between the locomotive and its tender. The tender was provided with a propane container which was made from solid drawn copper tube, silver soldered together with suitable staying so as to safely stand pressures up to 80 p.s.i. The container was also fitted with a needle control valve and a combined filler and non-return valve which was obtained from the manufacturers mentioned previously. In this way, the container could be filled from a large commercial gas bottle.

The method of propane firing which is described proved highly successful, but great care had to be taken when re-filling the container. Probably the safest method for small gauge locomotives is to use a comparatively small gas bottle obtainable from stores under the name of Camping-gaz, the filler and non-return valve being adapted according to the type of gas container.

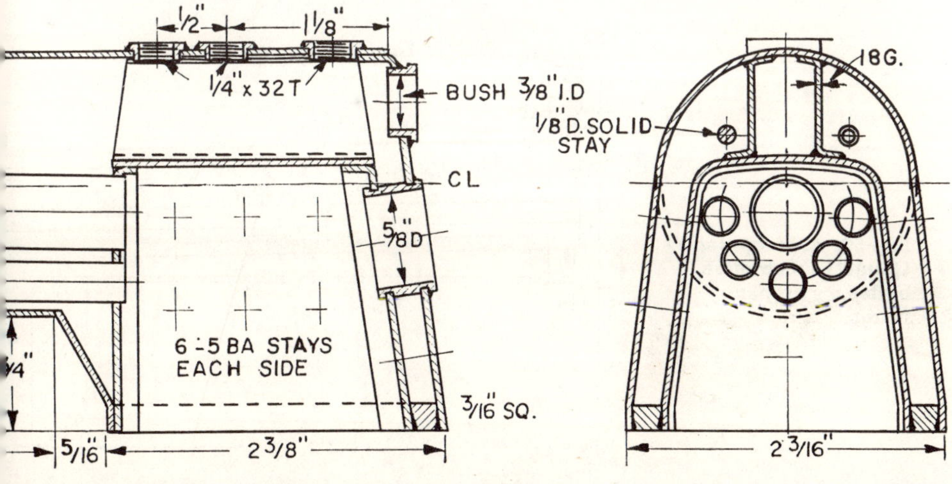

In the larger scales of model steam locomotives, the propane burner would have to be arranged in the form of two or four separate burner tubes and jets and they should be fitted to the supply tube in such a way that the flames produced from them are as diffuse as possible. It is also very important that the flames from such burners do not impinge directly on the ends of the tubes but on firebrick or other similar material arranged at an angle to the firebox tubeplate somewhat similar to the brick arch employed in full-size coal fired boilers.

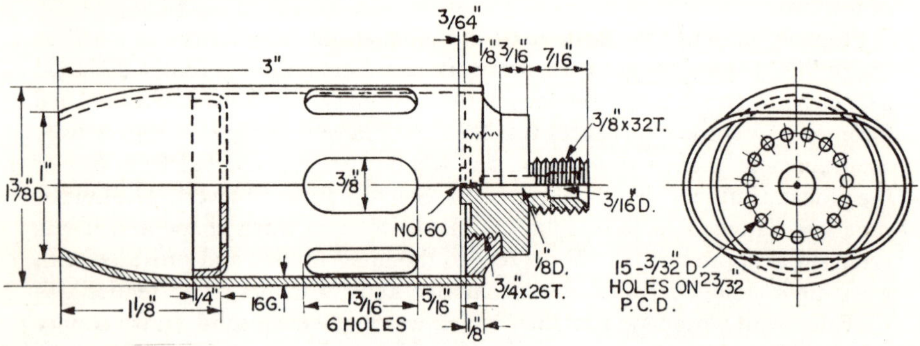

Fig. 49. A suggested design for a propane burner for 3½ in. gauge locomotives

Boilers for Propane Firing

Although water tube boilers may be employed for locomotives fitted with propane firing, the author believes that the best results are obtained by adopting the normal locomotive type of boiler, the firetubes being made considerably smaller in diameter than would be used for coal firing. This can be done for the principal reason that unlike the coal fired boiler, the propane fired boiler does not suffer from the building up of solid matter, cinders, etc., in the tubes.

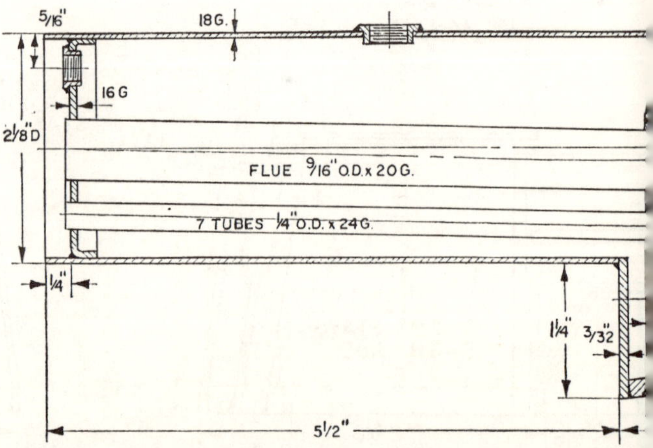

Fig. 50. A design for a Gauge 1 locomotive boiler for propane firing

Where the boiler required is of the large wide-firebox type, no combustion chamber should be provided, but the tubes should not be unnecessarily long otherwise the front end of the boiler will not be maintained at a temperature condusive to steam production. The normal firedoor may still be fitted and this should be provided with a means of adjusting the amount of air passing through it. The bottom of the firebox should provide for sufficient space for enough air to pass through to maintain efficient combustion, but the exact amount must be left to the builder as a matter of experiment.

Boilers for Spirit Firing

Even in the very smallest gauges, the water tube boiler is to be preferred to the simple "pot" boiler as it is a much quicker producer of steam, and to ensure success with methylated spirit firing, the material of the boiler should be as light as possible, consistent with safety. Seamless tube should be used for the inner barrel and also for the water tubes and these should be arranged as shown in Fig. 44 and silver soldered at both ends. It is a mistake to try to cram in too many tubes into the available space as methylated spirit firing does require plenty of ventilation for best results.

Stay-less Boilers

As the reader will appreciate, there is a great deal of labour involved in fitting stays to the conventional locomotive type boiler, for, as has been explained, all flat surfaces in such boilers have to be stayed unless the material itself is made exceptionally thick. It is, therefore, hardly surprising that attempts have been made, both in full-size locomotive work and in model engineering, to design a coal fired boiler for a locomotive without any stays, or at least without any of the short stays, which are normally fitted in the sides front and back of the firebox.

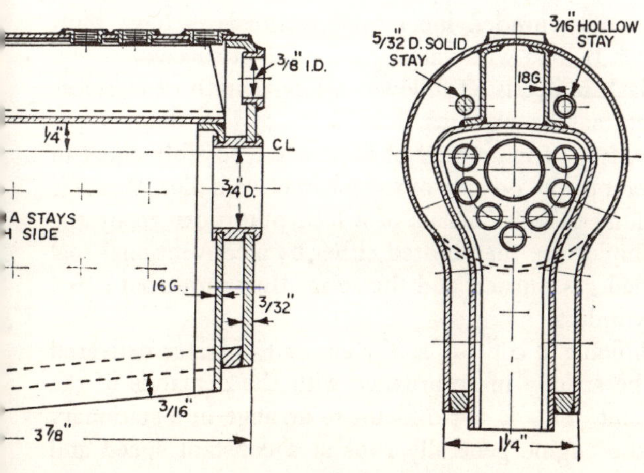

Fig. 51. This design is not recommended for coal firing

One solution to the problem is to make the firebox completely circular, with the grate arranged at some distance from the bottom, the air being admitted at the rear end. The firebox may also be heavily corrugated. An example of this in full-size locomotives was the special boiler fitted to some of the 0-8-0's of the old Lancashire and Yorkshire Railway. The author understands, however, that although these were cheap to build, they were not particularly satisfactory in service.

Another suggestion is to adopt the normal locomotive type barrel and instead of fitting the usual firebox with its flat plates, one or two circular drums of diameter equal to about a quarter of the main barrel, are fitted to the rear top end of the barrel longitudinally. From these drums, water tubes of quite small diameter are taken down at a slight angle to further water drums arranged on each side of the firebed, with additional tubes connecting their front ends to the bottom of the main barrel.

Cleaning plugs would have to be fitted to the ends of the water drums and provision would have to be made for bushes along the top of the upper drums for fitting the various safety valves, steam valves and so forth.

It would almost certainly be desirable to wrap asbestos sheet all around the water drums and their small tubes, but even so some difficulty might be caused through particles from the fire getting lodged in between the small tubes or between the tube and the outer casing.

Another problem would be how to fit a water gauge. It would not be possible to fit a long glass between one of the upper water drums and one of the lower drums unless the glass was very well protected from accidental damage.

Unfortunately, the author is not able to quote any performance figures with stay-less boilers of this type, though he understands that several have been made in the U.S.A.

Flash Boilers

One or two experimentally-minded locomotive enthusiasts have built locomotives fitted with flash boilers with varying amounts of success.

The principle of the flash boiler is as follows: a long length of stainless steel or copper tube, which may be anything from $\frac{3}{16}$ in. to $\frac{3}{8}$ in. diameter, is wound in the form of one, two or more coils of a diameter possibly equal to about one-third the diameter of the outside of the locomotive boiler. Water is fed into one end of this long tube by means of a feed pump driven by the engine. The coils of the boiler are then heated either by a conventional coal fire or by paraffin or bottled gas burner, and the steam thus produced is fed via the regulator to the cylinders.

The system is rather difficult to control, as the amount of water delivered to the boiler coils must be strictly in accordance with the demands of the engine at any given moment. This is not difficult to arrange in a stationary or marine plant, where the engine generally runs at a constant speed and

Plate 36

Mr. Ewins' 5 in. gauge 0–6–2 T. locomotive undergoing boiler tests

Below: propane firing applied to $3\frac{1}{2}$ in. gauge locomotive by F. Grosse-Holtfort of Hamburg

Plate 37

Cab view of Mr. F. Grosse-Holtfort's $3\frac{1}{2}$ in. gauge gas-fired model locomotive

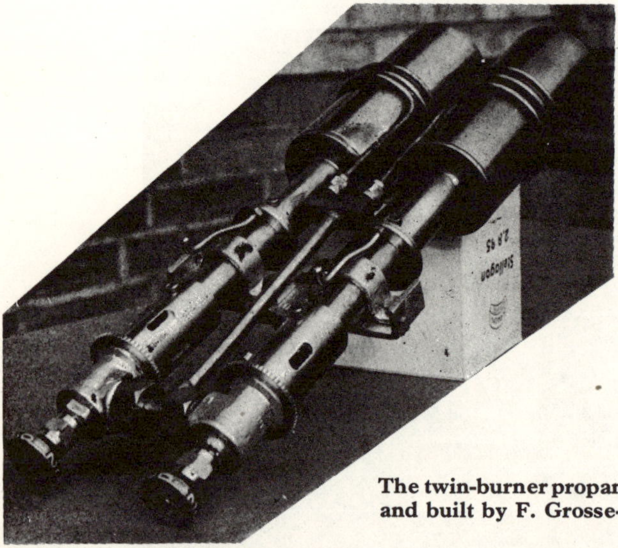

The twin-burner propane firing unit designed and built by F. Grosse-Holtfort of Hamburg

constant cut-off, but is very difficult on a locomotive, where regulator opening and cut-off have to be altered continuously, according to the nature of the road etc.

Lubrication is always more critical with flash steam, owing to the possibility of wide fluctuations in the temperature of the steam. A further complication is the greater quantity of scale, from the water, being passed through to the engine, with the possibility of damage resulting to valves and port faces.

It is desirable to incorporate some form of limiting valve or safety valve in a flash system, to avoid the danger of the pressure in the tubes rising momentarily to dangerous heights.

All in all, the flash system is not really suitable for locomotives, though it presents some interesting problems to the more experienced steam enthusiast prepared to spend a great deal of time (and possibly money!) on experimental work.

CHAPTER NINE

The Testing of Model Boilers

IT IS surprising how few model engineers have attempted any scientific tests on their model locomotive boilers.

Through the kindness of Mr. J. Ewins, the author is able to describe here some scientific tests on Mr. Ewins' 5 in. gauge coal-fired locomotive.

The locomotive concerned was a 0-6-2 tank engine, with inside cylinders $1\frac{5}{8}$ in. bore and $2\frac{1}{4}$ in. stroke; the valve gear was of the Joy type, with 80 per cent cut-off in full gear. Other particulars of the engine were as follows:

Grate area 21 sq. in.; fire tubes 16 in number, 12 in. long, $\frac{7}{16}$ in. bore; superheater flues 4 in number, $\frac{3}{4}$ in. bore; superheater elements made in stainless steel $\frac{1}{4}$ in., 22 gauge and projecting 7 in. into the firebox; driving wheel diameter $5\frac{1}{2}$ in.; blast nozzle 0·3 in. diameter.

A large snifting valve was fitted.

Tests made were concerned mainly with the boiler, but measurements were also made of power output and exhaust conditions, so as to correlate these with the performance of the boiler. Tests were made to establish the proportion of heat transferred by the tubes and the distribution of temperature at various points while the engine was working under different loads. Conditions of draught were investigated to decide the proportion of energy absorbed in drawing air and the products of combustion through the firebed and fire tubes respectively.

The temperature and pressure of the steam delivered to the cylinders was studied to determine the degree of superheat. Back-pressure at the exhaust nozzle was measured and correlated with power output. The overall thermal efficiency was also determined.

As originally built, the locomotive weighed 130 lb. in working order, but this was later increased to 180 lb. by the addition of ballast to improve adhesion. In service, the engine makes a surplus of steam on moderate and heavy loads, but with some coal needs a touch of the blower on very light loads. The main characteristics of this engine are its free running with regulator closed, its high maximum speed under load and its low water consumption.

The power output was measured by greasing the rails and allowing the wheels to slip while the drawbar pull was measured on a Salter balance and the speed by Smith tachometer. Various loads were obtained by taking some of the engine weight on ball bearing mounted wheels, which were adjustable.

The friction of these wheels against the rails was small and was therefore ignored.

Measurement Technique

It was not possible to measure directly the rate at which the fuel was used over short test periods, as the firebox contents could not be accurately known. If the rate of flow of the flue gas was measured together with its chemical composition, it was possible to deduce the rate at which the fuel was being used and heat generated. The difficulty here lay in measuring the rate of flow of the flue gas from the chimney when it was mixed with the exhaust steam. To overcome this, the flue gas and the exhaust steam were passed through a water-cooled baffle, caught in a large polythene bag and the time noted for this to be filled. The volume of gas caught was then measured by a water displacement method.

A slight error can occur with this technique, because any water contained in the flue gas is also condensed along with the steam and therefore not measured. However, by using anthracite as fuel, little water vapour is produced and the error is very small.

The smokebox and firebox vacua were measured by water manometers, while the back-pressure at the exhaust nozzle was indicated on a mercury manometer.

Temperatures were measured by inserting Chromel-Alumel thermocouples at the various points and bringing these out to a 12-way thermocouple switch, using a common cold junction and meter. In the case of the firebed temperature, a platinum/platinum-rhodium couple was used in a protective sheath and connected to a separate meter.

Steam chest pressure was measured by a 0 to 100 lb. per sq. in. Bourdon tube gauge, the calibration of which was checked. The temperature of the steam delivered to the steam chest was measured by screwing a small copper thimble containing a thermocouple into the steam pipe close to it. The thimble was pointed to project inwards about two-thirds of the pipe diameter and was made very thin.

The exhaust temperature thermocouple was inserted directly into the side of the exhaust nozzle and the exhaust back-pressure manometer connected to a point in the exhaust pipe about 1 in. from the nozzle.

The thermocouple used to measure the firebed temperature was inserted through a hole in a plate covering the firehole.

Flue gas samples were collected in plastic containers that had previously been used to contain washing-up liquid. These bottles were provided with tightly fitting rubber stoppers, through which glass tubes were passed. Connection between these tubes and the sampling tube was made by means of a short rubber connector with pinch-clip. The sampling tube was $\frac{1}{8}$ in. diameter copper, pointed to slide through one of the fire tubes from the smokebox end and marked to indicate when its open end coincided with each end of the fire tube. Each sample bottle was evacuated, whereupon it collapsed.

When connected to the sampling tube and the pinch-clip released, the gas was drawn gently in, until the bottle resumed its normal shape. This procedure proved very convenient, both in taking the sample and subsequently when the analysis was conducted by means of an Orsat apparatus. The reason for taking samples from both ends of the fire tubes was to check whether or not combustion took place within these tubes. The tests indicated no combustion.

In order to simplify the thermal calculations, anthracite fuel was used, since this contains about 90 per cent carbon, 4 per cent ash and moisture, whilst the 6 per cent residue is only partially combustible material. The error in assuming that only the 90 per cent carbon takes part in the combustion is quite small.

When carbon burns in air, the oxygen contained in the air combines with it to form carbon dioxide, the reaction being accompanied by the liberation of heat. The chemical reaction for 1 lb. of carbon can be written:

$C + 2\,O = CO_2$ 14,600 B.T.U. of heat.
The proportions by weight are:
12 parts carbon + 32 parts oxygen = 44 parts carbon dioxide.
Therefore, 1 lb. of CO_2 indicates the production of 4,000 B.T.U.

The Orsat apparatus used for the analysis was arranged to determine the parts by volume of carbon dioxide, carbon monoxide and oxygen in the flue gas. The presence of carbon monoxide indicates incomplete combustion, whilst oxygen shows that excess air was being supplied. The amount of carbon dioxide indicates the extent of combustion and the heat generated, the number of B.T.U. being obtained from the relationship given above. To do this however, the mass of carbon dioxide must be found.

The volume rate of flow of flue gas having been determined by the use of the bag, it is only necessary to multiply the percentage of this volume due to carbon dioxide by the density of carbon dioxide (0·12 lb./ft.^{3}) to obtain the mass flow of CO_2. Finally, as we have seen, the mass flow of CO_2 multiplied by 4,000, gives the rate of heat liberation in B.T.U./min. if the original volume was measured in ft.^{3}/min.

The proportion of this heat transferred in the fire tubes may be obtained, from a knowledge of the temperature fall of the flue gas during its passage along the tubes. In order to do this, the specific heat and constant pressure of the gas must be known and since this is a mixture of gases there is some difficulty. Fortunately, the major gases present are carbon dioxide and the constituents of air. The former has a specific heat of 0·201 and the latter 0·237. If separate calculations are made for these two components and the results added, there will be no great error.

Knowing the heat transferred in the tubes and the total heat evolved, both the proportion transferred in the tubes and that rejected at the smokebox can now be deduced. This last figure enables one to calculate the boiler efficiency

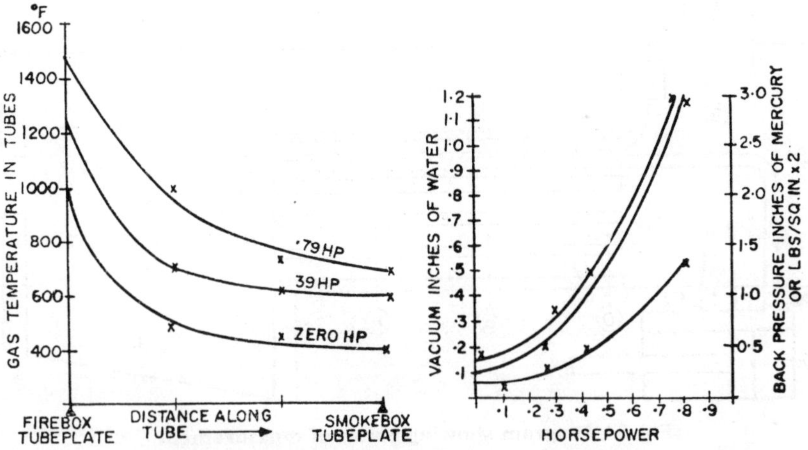

Fig. 52. Graphs showing gas temperatures, vacuum etc.

except that it will not take into account loss by radiation and unburnt fuel, ejected from the chimney.

Most of the tests were carried out at a speed corresponding to 7 m.p.h. (400 r.p.m.), the measurements being taken after the engine had been running sufficiently long for conditions to have become steady, with the boiler pressure at 80 lb. Owing to the resistance of the water-cooled baffle, it was found that under certain loads there was a tendency for the pressure to fall, unless the blower was used. It was therefore decided to limit this part of the test to three outputs of approximately $\frac{1}{2}$ h.p., $\frac{1}{8}$ h.p. and zero output (driving wheels suspended) and to follow these with a series of tests at various outputs, ranging from zero to 0·8 h.p. without the baffle.

Results

The observations of the first series of three tests are recorded on the accompanying diagram, which shows also the results of calculations using the values obtained. The results from the second series of tests have been combined with some of those from the first series and presented on the graphs of Fig. 54.

Different aspects of the engine's operation will now be considered in the light of these results.

Heat Transfer

It will be seen that the porportion of heat transferred in the tubes varies with the flue gas composition, which in turn varies with the load. In test one, with the very light load, there was an excess of oxygen, whereas in test two the combustion was better, 13 per cent CO_2 being produced with low excess oxygen. This caused an intense fire (not cooled by a large excess of air),

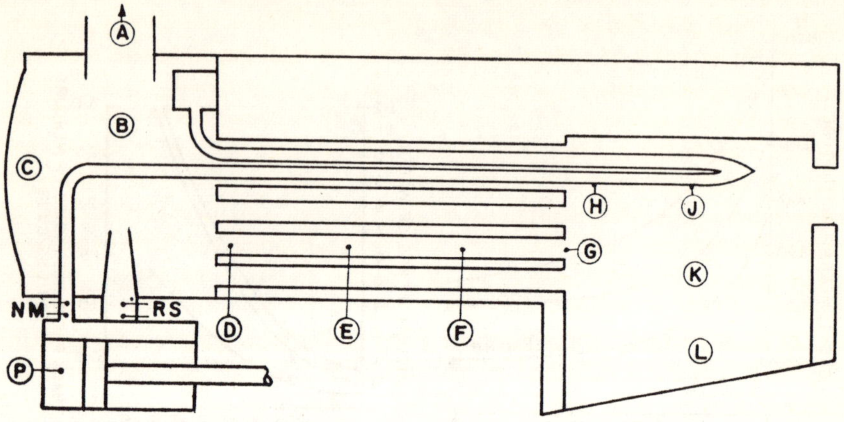

Fig. 53. Diagram showing points of measurement

		Test No. 1 (no load)
Speed		400 r.p.m. (6·5 m.p.h.)
Tractive Force		Zero lb.
D.B.H.P.		Zero
Heat Generated		290 B.T.U./min. 100%
Heat Transferred in Firetubes		74 ,, ,, 25%
Heat Rejected to Smokebox		38 ,, ,, 13%
Heat Transferred in Firebox (by difference)		178 ,, ,, 62%
Overall Thermal Efficiency		Zero %
Coal Consumption		—lb./D.B.H.P.—hr.
Grate Loading		—D.B.H.P./sq. ft.
Rate of Flow of Flue Gas Recorded at	A	6·7 cubic ft./minute
Flue Gas Analysis ,, ,,	B	CO_2 10%, O_2 9%, CO 0·5%
Smokebox Vacuum ,, ,,	C	0·3 in. water
Temperature in Firetube ,, ,,	D	360 F.
,, ,, ,, ,, ,,	E	490 F.
,, ,, ,, ,, ,,	F	550 F.
,, ,, ,, ,, ,,	G	950 F.
Superheater Temperature ,, ,,	H	800 F.
,, ,, ,, ,,	J	850 F.
Firebox Vacuum ,, ,,	K	0·05 in. water
Fire Bed Temperature ,, ,,	L	2120 F.
Steam Chest Pressure ,, ,,	M	6 lb./in. sq.
Steam Temperature ,, ,,	N	500 F.
Cylinder Temperature ,, ,,	P	250 F.
Exhaust Back Pressure ,, ,,	R	0·18 lb./sq. in.
Exhaust Temperature ,, ,,	S	310 F.
Superheat		270 F.

which in turn gave rise to the high transfer in the firebox of 320 B.T.U./min. Also, in the absence of a large excess of air, the total gas flow was relatively low, resulting in a relatively smaller proportion of the heat being carried in the flue gas, a boiler efficiency of 85 per cent, which is comparable with full-size practice. Under the heavier load of test three, a large value of excess oxygen appeared again, giving the lower percentage transfer in the firebox and higher in the tubes, with efficiency falling to 78 per cent. This is evidence that the fire was being overdriven, producing poorer combustion and little extra heat.

The graphs of Figure 54 indicate the manner in which the temperature of the flue gas falls as it passes along the tubes. It will be noticed that generally the fall is very rapid for the first one-third of the tube length, although this is not so pronounced with the heavier output. At moderate loading, more than 80 per cent of the heat transfer in the tubes takes place in the first one-third of their length. In fact, the last one-third contributes practically nothing and

Test No. 2	Test No. 3
400 r.p.m. (6·5 m.p.h.)	400 r.p.m. (6·5 m.p.h.)
7 lb.	30 lb.
0·125	0·525
445 B.T.U./min. 100%	455 B.T.U./min. 100%
83 ,, ,, 19%	124 ,, ,, 19%
42 ,, ,, 9%	77 ,, ,, 17%
320 ,, ,, 72%	254 ,, ,, 56%
1·19%	4·9%
15 lb./D.B.H.P.—hr.	3·9 lb./D.B.H.P.–hr.
0·9 D.B.H.P./sq. ft.	3·8 D.B.H.P./sq. ft.
7·28 cubic ft./minute	10·8 cubic ft./minute
CO_2 13·5%, O_2 3·5%, CO ·5%	CO_2 9%, O_2 10%, CO 1%
0·35 in. water	0·6 in. water
370 F.	430 F.
410 F.	690 F.
570 F.	760 F.
980 F.	1070 F.
980 F.	1570 F.
1570 F.	1460 F.
0·15 in. water	0·3 in. water
2840 F.	2912 F.
11.5 lb/in. sq.	41 lb./in. sq.
620 F.	595 F.
275 F.	295 F.
0·22 lb./sq. in.	0·9 lb./sq. in.
325 F.	350 F.
378 F.	307 F.

the last inch or two only just about makes up for the loss by radiation of the exterior of the boiler barrel surrounding this part of the tube bank. It is only at very heavy loading that the smokebox ends of the tubes make a positive contribution and this in a boiler which has a short barrel by normal standards.

Superheat

The figures for superheat are shown for the three tests as 270 deg. F., 320 deg. and 307 deg. respectively. They are somewhat greater than normal in full size working, where about 250 deg. F. of superheat, i.e. a steam temperature of 650 deg. F. for a boiler pressure of 220 p.s.i. are obtained.

The cylinders of the model were cast in phosphor-bronze and the pistons made in drawn bronze. Recently the pistons were withdrawn for the purpose of replacing the graphite packing and in spite of the high superheat and nearly 1,000 actual miles of heavy use, the pistons and bores were found to be in excellent condition. Shell Valvata J78 has always been used for cylinder lubrication, fed by mechanical lubricator.

Exhaust steam temperatures are seen to be very high, showing a considerable degree of superheat. Under these conditions it is quite certain that no cylinder condensation takes place. Indeed, the cylinder wall temperature was, in all the tests above, the saturation temperature of the steam in the steam chest.

Exhaust Back-pressure

As stated earlier, the back-pressure was measured just behind the exhaust nozzle and as the results show never exceeded 1·5 p.s.i. Under these circumstances there was no point in considering possible improvements in blast pipe arrangements, since any gain here would have had a very small effect on the working of the engine. The reason for such low values becomes apparent when one considers the flow of gas through the firetubes. From the results of test three, the velocity of the gas was found to be 570 ft. per minute in tubes of $\frac{7}{16}$ in. bore. Under these circumstances, the flow is streamline and as a result the pressure difference across the tubes (smokebox vacuum minus firebox vacuum) to cause this flow is small (0·35 in. of water). In full size practice the gas flow in the tubes is turbulent and therefore needs a much greater pressure difference, figures of 5 in. of water and 10 lb. back-pressure being obtained.

Firebed Temperature

Here the values obtained were considerably greater than in full size working and it is probably here that the model makes up for some of the disadvantage it suffers relative to full size elsewhere. The problem of clinker formation at high temperatures is not so acute in models, becuase they are rarely in steam for such long periods as in full size practice the elevated temperature of 2,900 deg. F. compared with about 2,500 deg. F. in full size greatly facilitates heat transfer in the firebox by radiation.

In test three the engine was well loaded and the coal consumption figure of 3·9 lb./D.B.H.P./hour is very similar to that obtained in full size. Under these conditions the grate loading of the model was 3·8 D.B.H.P./ft.^{2}, whereas in full size a value of about 30 D.B.H.P./ft.^{2} would be usual. If the fire of the model had been of scale thickness and the coal burning at the same rate per unit volume as in the full size engine, then probably the same temperature would have been attained. The temperature was, however greater in the model, thus indicating a greater rate of combustion and the D.B.H.P./ft.^{2}, instead of being "scale", i.e. 30/12 = 2·5 D.B.H.P./ft.^{2}, was 3·8 D.B.H.P./ft.^{2}, which is greater, as expected.

Conclusions

In general it appears that the operating characteristics of the model boiler are very similar to those of a full-size boiler. This comes about by compensating scale effects. For instance, heat transfer in the tubes in full size is greatly aided by turbulence; whereas in the model, where the flow is streamline, the tubes are of much smaller diameter, resulting in reduced resistance to the radial flow of heat, which thus compensates. If the full-size engine were provided with the same size tubes as the model, better heat transfer would occur, but the tubes would choke up very quickly. Why do they not choke up quickly on the model? Because the gas velocity is roughly scale, i.e. $\frac{1}{12}$, and therefore pieces of coal sufficiently large to choke the $\frac{7}{16}$ in. tubes are not lifted from the fire. Some coal is lifted, but only that which is finer and able to pass through without choking.

Again the model is at a disadvantage, due to its radiating surface being proportionately larger compared with the volume it encloses than in full size. With most models this results in severe condensation in the cylinders, but as the results show, this is readily offset by super-heating the steam, provided a radiant superheater is employed, thus taking advantage of the propitious conditions in the firebox. The model is also favoured by the natural laws of gas flow, resulting in exhaust back-pressure being extremely small. At the beginning of this chapter a number of questions were posed. Which of these can now be answered?

The answer to the question of tube length seems fairly clear: it would seem difficult to make it too short. Even by halving the length—a pretty drastic measure—little would be lost in heat transfer. Drastically cutting down the barrel length in this way would adversely affect the free water surface, from which the steam is released into the steam space and cause wet steaming; but it is doubtful if trouble of this sort would be met if, say, the tubes had been reduced by one-third of their length, because the rate of steam production in this last third is very small and therefore the surface above this section is made little use of for steam release.

What is to be gained by cutting down the barrel and tube length? Before conducting these tests, Mr Ewins thought that the chief gain would be a

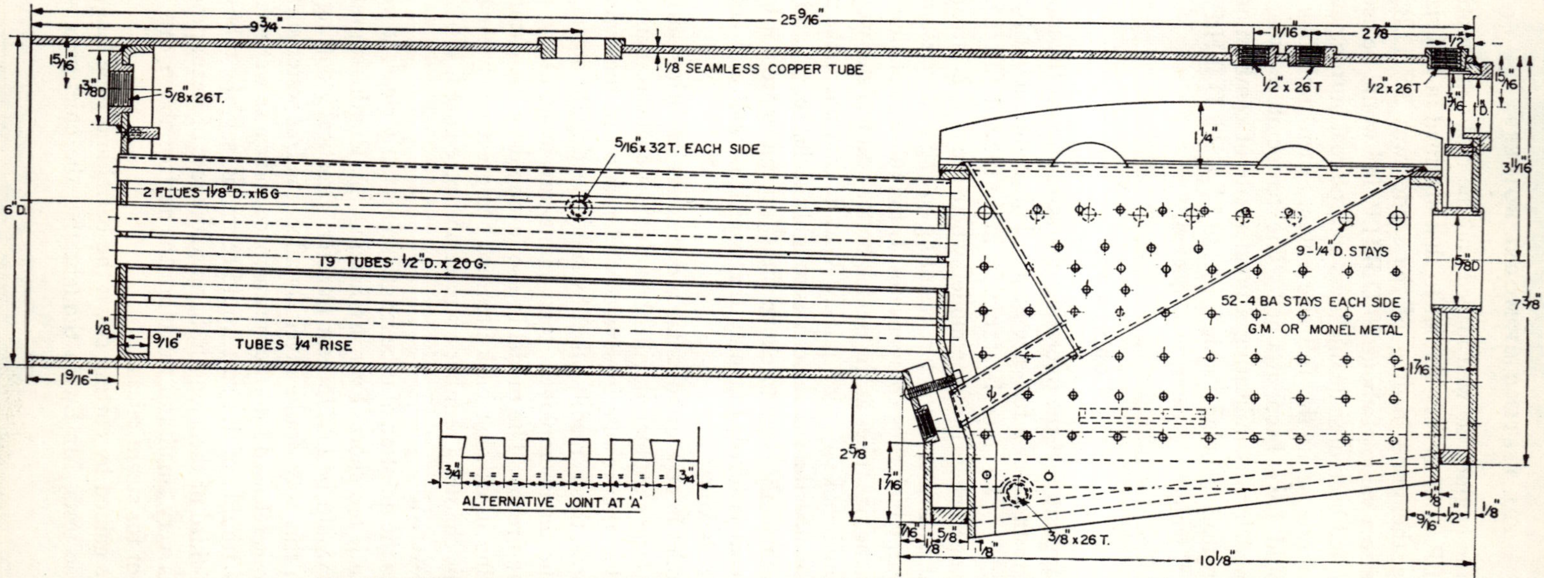

Fig. 54. The author's latest boiler design: a large roundtop boiler for a 5 in. gauge G.N.R 2-8-0. Note the thermic syphon and large water space at the throatplate

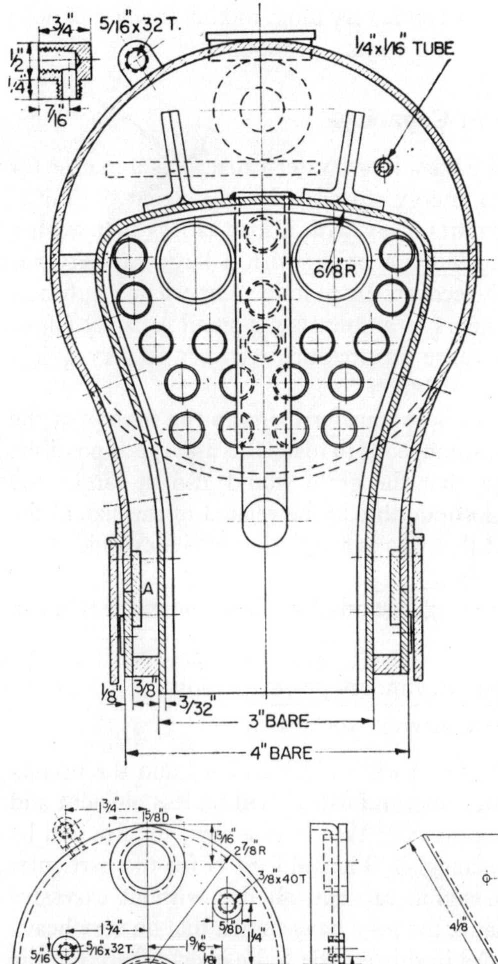

Fig. 55. Cross-section of the boiler shown in Fig. 54

Below: Fig. 56. Details of the backhead and the thermic syphon

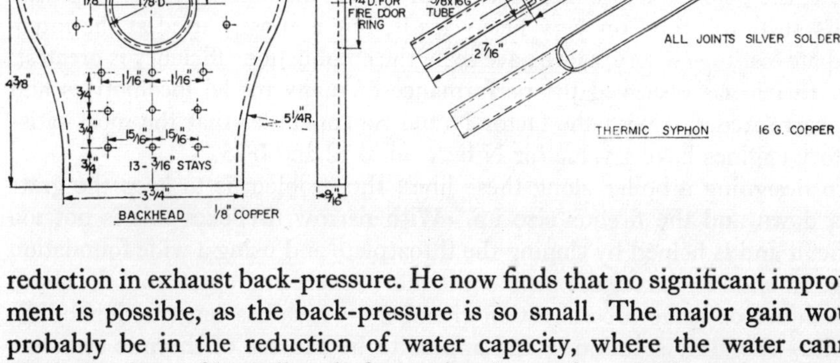

reduction in exhaust back-pressure. He now finds that no significant improvement is possible, as the back-pressure is so small. The major gain would probably be in the reduction of water capacity, where the water cannot obtain its share of heat directly from a heating surface. Such water must gain most of its heat by mixing with water from elsewhere, or by condensing steam from the steam space above. If the barrel is very long, mixing (by circulation) becomes less effective and the latter effect more pronounced.

Water Capacity

Whether or not a boiler should have a large water capacity is a matter for debate. With a large capacity, the energy stored is large, so that the initial fall in pressure when the engine starts away with a load is less than with a small capacity. On the other hand, once a boiler with a large capacity has dropped in pressure, it is slow to recover. In practice, experience indicates that, provided an engine has cylinders of ample size, a small capacity (short barrel) boiler is advantageous, because its pressure recovery is very quick, due to all its heating surface being effective.

In view of the high proportion of heat transfer being in the firebox of the engine under test, it would seem advantageous to make this as large as possible. There is a danger here, however, that the grate would also be large—too large in fact. The grate area should undoubtedly be related to the size in the engine part of the locomotive and that this is a function of the cylinder bore and stroke and driving wheel diameter.

Mr. Ewins has devised the following formula by which the grate area can be related to the engine.

$$N = \frac{\text{Swept volume of cylinders per revolution}}{\text{Grate area} \times \text{driving wheel diameter}}$$

In engines where N is high, the grate will be driven hard and the firebox heat transfer will be more efficient, but combustion will be less efficient and the tendency to clinker more pronounced. Where N is low, the fire will be dull and the boiler inefficient and sluggish. The value of N for the particular engine tested was 0·15 and this engine can run all day without excessive clinker formation. On the other hand, the tests have shown that on very heavy loads there is a tendency for the fire in this engine to be over-driven. On the whole, the position is probably satisfactory, since an engine spends very little of its steaming time on very heavy loads, but is mostly used at the intermediate loadings where, as we have seen, the combustion efficiency is greatest. Mr. Ewins has observed the performance of many model locomotives and has correlated this with the factor N and he concludes that the most satisfactory engines have a value for N between 0·12 and 0·25.

In designing a boiler along these lines, the problem is to keep the grate area down and the firebox size up. With narrow fireboxes this is not too difficult and is helped by sloping the throatplate and using a wide foundation ring. In wide fireboxes the problem is more severe and apart from the above techniques or using a combustion chamber—which has constructional difficulties—the only solution would appear to be the use of thermic syphons. Mr. Ewins' conclusions, therefore, would seem here to agree with the author's.

Two such thermic syphons will effectively divide a wide firebox into three narrow ones, thus the heating surface is increased where it is most effective and better circulation is obtained.

The question of superheat has been answered, but only for an engine having a radiant type superheater. For the conventional type of superheater, the results of this test suggest that it would not be very effective. The so-called "hairpin" superheater usually ends about an inch or two from the firebox, where we have seen the temperature is well down on that in the firebox proper and no radiant transfer of heat is possible.

As regards blastpipes, it would seem that no great improvement can be expected by using more elaborate types.

Finally, the question of the utilisation of the heat delivered to the smokebox. The tests show that this is about 13 per cent of the heat generated and that the smokebox temperature is usually about 400 deg. F. An estimate of the heat lost by convection and radiation from the cylinders gives a value of about 6 B.T.U./min. when these are maintained at the saturation temperature of steam at 80 p.s.i. This heat loss is about one sixth of the heat value rejected in the flue gases. The difficulty lies in inducing this heat to flow to the cylinders with a temperature difference of only 75 deg. F. For a time it was thought that this could be achieved by providing fins in the smokebox and heavy copper lugs on the outside, reaching down to the cylinders. If, however, one considers the heat transfer to these fins and assumes it to be comparable with that taking place in the last few inches of the firetubes, it soon becomes apparent that the scheme is hardly worthwhile.

An alternative to the above is to fit a feed water heater in the smokebox, where heat transfer would be good owing to the large temperature difference between the water and the flue gases. For this to be effective the water would need to be cold and therefore delivered by mechanical pumps. But the simplicity of injectors is to be preferred and in many cases additional complication for the sake of a small gain in efficiency is not worthwhile. Efficiency is relatively unimportant in model engineering; reliability and functional convenience are much more desirable.

In this connection, there would seem to be room for improvement in model locomotives as regards the initial condensation which takes place in the cylinders on starting away after a period of standing. High superheat helps a lot, but if the cylinders can be kept warm, as indicated above, the position would be better. However, little can be done in this direction. An alternative would be to make the thermal capacity of the cylinders low, so that they warm up quickly and to this end built-up cylinders would undoubtedly help. Surprise may be expressed on reading through the results of these tests, at the high percentage of heat transfer in the superheater. When calculated from the results, this turns out to be about 10 per cent of the heat generated. If the surface area on that part of the superheater in the firebox is compared with the internal area of the firebox, it is found to be about 15 per cent, so that it is not surprising that 10 per cent of the heat transfer takes place here. Whether this high degree of superheat is entirely justified is open to question, but the engine tested is very lively and the cylinders and valves have stood up very well to the high temperatures involved, so there would seem to be no obvious drawback functionally.

To sum up, the tests show that this boiler had an efficiency of about 80 per cent, so that no very great improvement in this respect could be expected by varying the design. There is no question that radiant superheaters can produce a high degree of superheat and this can be tolerated by cylinders made of non-ferrous material, providing that lubrication is good.

Tests in South Africa

Tests of model locomotive boiler efficiency have also been carried out by Mr. J. Busbridge of the Cape Town Society of Model Engineers, with the cooperation of technicians of the University of Cape Town. The boiler concerned in the tests was a $\frac{3}{4}$ in. scale wide firebox "Britannia" type, fitted with a combustion chamber and three superheater flues of 1 in. diameter. The fuel used was Welsh Steam Coal of 15,000 B.T.U./lb. approx.

In the first test, a firing rate of 0·925 lb. per hour was used, with a boiler working pressure of 73 p.s.i. The temperature of the superheated steam was 352 deg. F. and the overall efficiency of the boiler worked out at 70 per cent.

In the second test, at a firing rate of 1·725 lb./hr., the superheated steam temperature was 404 deg. F. and overall efficiency 58·5 per cent.

In the third test, at a firing rate of 1·8 lb./hr., sup. steam temperature was 444 deg. F., and efficiency 77 per cent.

In the fourth test, at a firing rate of 2 lb./hr., sup. steam temperature was 455 deg. F., and efficiency 69·5 per cent. and in a final test, at a firing rate of 2·83 lb./hr., the superheated steam temperature was 509 deg. F., and overall efficiency 65 per cent.

In the last test, water evaporation was around $2\frac{1}{4}$ gallons per hour. It was concluded that the boiler was capable of an even higher rate of steam production, and that in spite of considerable losses of heat from the boiler at the chimney, by radiation and through unburt fuel, the effective heat still available amounted to about 77 per cent of the whole. As to the overall efficiency of the boiler, though the figures produced were a little lower than in Mr. Ewins' tests, they show that a well-designed model locomotive boiler is quite an efficient steam raiser, giving figures which compare closely with the best full-size practice.